T0318554

Economic Evaluation in Genomic and Precision Medicine

Translational and Applied Genomics Series

Economic Evaluation in Genomic and Precision Medicine

Series Editor: **George P. Patrinos**

Edited by

Christina Mitropoulou

The Golden Helix Foundation, London, United Kingdom; Department of Genetics and Genomics, College of Medicine and Health Sciences, United Arab Emirates University, Al-Ain, United Arab Emirates

Sarah Wordsworth

Health Economics Research Centre, Nuffield Department of Population Health, University of Oxford, Oxford, United Kingdom; Oxford National Institute for Health Research Biomedical Research Centre, Oxford, United Kingdom

James Buchanan

Health Economics Research Centre, Nuffield Department of Population Health, University of Oxford, Oxford, United Kingdom; Oxford National Institute for Health Research Biomedical Research Centre, Oxford, United Kingdom

George P. Patrinos

Department of Pharmacy, School of Health Sciences, University of Patras, Patras, Greece; Department of Genetics and Genomics, College of Medicine and Health Sciences, United Arab Emirates University, Al-Ain, United Arab Emirates; Zayed Center of Health Sciences, United Arab Emirates University, Al-Ain, United Arab Emirates

Academic Press is an imprint of Elsevier
125 London Wall, London EC2Y 5AS, United Kingdom
525 B Street, Suite 1650, San Diego, CA 92101, United States
50 Hampshire Street, 5th Floor, Cambridge, MA 02139, United States
The Boulevard, Langford Lane, Kidlington, Oxford OX5 1GB, United Kingdom

ISBN: 978-0-12-813382-8

For information on all Academic Press publications
visit our website at https://www.elsevier.com/books-and-journals

Publisher: Stacy Masucci
Acquisitions Editor: Peter B. Linsley
Editorial Project Manager: Kristi Anderson
Production Project Manager: Punithavathy Govindaradjane
Cover Designer: Mark Rogers

Typeset by STRAIVE, India

Contents

Contributors xi
Preface xiii

CHAPTER 1 Introduction to economic evaluation in health sciences 1
Christina Mitropoulou, Vasileios Fragoulakis, James Buchanan, Sarah Wordsworth, and George P. Patrinos

1.1 Introduction.. 1
1.2 Key economic concepts and terminology used in health economics.................... 2
1.2.1 Budget constraints and demand.. 2
1.2.2 Utility.. 4
1.2.3 Indifference curves.. 5
1.2.4 Social welfare.. 7
1.3 Economic evaluation.. 9
1.3.1 Rationale for economic evaluation.. 9
1.3.2 Types of economic evaluation.. 10
1.4 Concluding remarks.. 12
References.. 13

CHAPTER 2 Foundations of pharmacogenomics and personalized medicine 15
Maria Koromina and George P. Patrinos

2.1 Introduction.. 15
2.2 Implementing pharmacogenomics into clinical care.. 16
2.2.1 Pharmacogenomics for cancer therapeutics.. 16
2.2.2 Pharmacogenomics for drug treatment of cardiovascular diseases 17
2.2.3 Pharmacogenomics for psychiatric diseases.. 18
2.3 Electronic tools for translating genomic findings into a clinically meaningful format 19
2.4 Assessing the ethical, societal, and financial aspects of personalized medicine.. 22
2.4.1 Ethical and legal issues.. 23
2.4.2 Raising genomics awareness among healthcare professionals and the general public.. 23
2.4.3 Economic evaluation of personalized medicine interventions................. 24
2.5 Large-scale personalized medicine implementation efforts worldwide 24

2.6 Conclusions and future perspectives 25
Acknowledgments 26
References 26

CHAPTER 3 Economic evaluation of genome-guided treatment in oncology.... 33
Paula K. Lorgelly
3.1 Introduction 33
3.2 Methodological challenges 35
3.3 The inseparability of test and treatment 35
3.4 Value of testing 37
3.5 Cost-effectiveness comparator 38
3.6 Pricing and reimbursement 39
3.7 Conclusion 42
References 43

CHAPTER 4 Economic evaluation of rare diseases and the diagnostic odyssey 47
Dean A. Regier, Deirdre Weymann, Ian Cromwell, Morgan Ehman, and Samantha Pollard
4.1 Introduction 47
4.2 Genomics and rare diseases 47
4.2.1 Diagnostic odyssey 48
4.2.2 Opportunity for genomics 48
4.3 Economic evaluations of genomics in rare diseases 51
4.3.1 Step 1: Determine comparators and study design 52
4.3.2 Step 2: Estimate health and non-health outcomes 55
4.3.3 Step 3: Estimate costs 59
4.3.4 Step 4: Estimate relationship between costs and outcomes 62
4.4 Conclusion 66
Acknowledgment 67
References 67

CHAPTER 5 Economic analysis of pharmacogenetics testing for human leukocyte antigen-based adverse drug reactions 71
Rika Yuliwulandari, Usa Chaikledkaew, Kinasih Prayuni, Hilyatuz Zahroh, Surakameth Mahasirimongkol, Saowalak Turongkaravee, Jiraphun Jittikoon, Sukanya Wattanapokayakit, and George P. Patrinos
5.1 Introduction 71
5.2 Economic evaluation of adverse drug reaction-related human leukocyte antigen genotyping, its guidelines, and current implementations 72

5.2.1 Abacavir-induced hypersensitivity reaction 72
5.2.2 Allopurinol-induced hypersensitivity reaction 76
5.2.3 Carbamazepine-induced hypersensitivity reaction 76
5.3 Conclusions 81
References 82

CHAPTER 6 Economic evaluation of personalized medicine interventions in medium- and low-income countries with poor proliferation of genomics and genetic testing 87
Christina Mitropoulou and George P. Patrinos
6.1 Introduction 87
6.2 Implementing personalized medicine interventions beyond high-income countries: Setting the scene 88
6.2.1 Challenges in low-income and lower middle-income countries outside Europe 89
6.2.2 Challenges in upper middle-income countries in Europe 89
6.3 Examples of economic evaluations of personalized medicine interventions in medium- and low-income countries 91
6.3.1 Anticoagulation and antiplatelet therapies 91
6.3.2 Antidepressants 92
6.3.3 Allopurinol 92
6.3.4 Human papillomavirus testing 93
6.4 A generic model as tool for measuring cost-effectiveness of personalized medicine interventions in medium- and low-income countries 94
6.5 Conclusion and future perspectives 95
Acknowledgments 96
References 96

CHAPTER 7 Theoretical models for economic evaluation in genomic and personalized medicine 99
Vasileios Fragoulakis, George P. Patrinos, and Christina Mitropoulou
7.1 Introduction 99
7.2 The genome economics model 102
7.3 Generalization of the genome economics model 106
7.4 Perspectives 110
References 111

CHAPTER 8 Using "big data" for economic evaluations in genomics 113
Sarah Wordsworth, Brett Doble, Katherine Payne, James Buchanan, Deborah Marshall, Christopher McCabe, Kathryn Philips, Patrick Fahr, and Dean A. Regier

8.1 Introduction 113
8.2 Using "big data" for the economic evaluation of genomic tests 114
8.3 Analytics for omics data 115
8.4 Difficulties in using "big data" in for economic evaluation of sequencing tests 115
8.4.1 Data collection challenges 116
8.4.2 Data management challenges 116
8.4.3 Data analysis challenges 118
8.4.4 Identifying key challenges 119
8.5 Conclusions 119
References 120
Further reading 121

CHAPTER 9 Assessing the stakeholder environment and views towards implementation of personalized medicine in a healthcare setting 123
Christina Mitropoulou, Athanassios Vozikis, and George P. Patrinos

9.1 Introduction 123
9.2 Identifying stakeholders in personalized medicine 124
9.3 Eliciting and analyzing stakeholder views and opinions related to personalized medicine 124
9.4 Implementing stakeholder analysis in genomic and personalized medicine: An example from the preliminary assessment of the genomic and personalized medicine environment in Greece 127
9.5 Defining opportunities and threats when implementing genomic and personalized medicine in Greece 131
9.6 Concluding remarks 132
Acknowledgments 133
References 133

CHAPTER 10 Feasibility for pricing, budget allocation, and reimbursement of personalized medicine interventions 135
Christina Mitropoulou, Margarita-Ioanna Koufaki, Athanassios Vozikis, and George P. Patrinos

10.1 Introduction 135
10.2 Institutions involved in pricing and reimbursement 137

10.3 Coverage, pricing, and reimbursement strategies for genomic testing services ... 138
10.4 A proposed strategy for pricing and reimbursement in personalized medicine ... 140
10.4.1 Ensure access to essential genomic testing services for all at acceptable and affordable prices for the healthcare system ... 140
10.4.2 Establish a common and centralized regulation to ensure safety, efficacy, quality, and fairness, while allowing space for innovation necessary to move the field forward ... 142
10.4.3 Implementation of genomic tests and information by physicians, according to patient needs and clinical utility/actionability of testing outcomes ... 142
10.4.4 Invest in the research of personalized medicine, evaluate novel and existing diagnostic procedures, and monitor patient safety ... 143
10.5 Restrictions and concerns ... 144
10.6 Conclusions ... 145
References ... 146

Index ... 149

Contributors

James Buchanan
Health Economics Research Centre, Nuffield Department of Population Health, University of Oxford, Oxford, United Kingdom; Oxford National Institute for Health Research Biomedical Research Centre, Oxford, United Kingdom

Usa Chaikledkaew
Faculty of Pharmacy, Mahidol University, Salaya, Thailand

Ian Cromwell
Cancer Control Research, BC Cancer, Vancouver, BC, Canada

Brett Doble
Paraexel, London, United Kingdom

Morgan Ehman
Cancer Control Research, BC Cancer, Vancouver, BC, Canada

Patrick Fahr
Health Economics Research Centre, Nuffield Department of Population Health, University of Oxford, Oxford, United Kingdom; Oxford National Institute for Health Research Biomedical Research Centre, Oxford, United Kingdom

Vasileios Fragoulakis
The Golden Helix Foundation, London, United Kingdom

Jiraphun Jittikoon
Faculty of Pharmacy, Mahidol University, Salaya, Thailand

Maria Koromina
Department of Pharmacy, School of Health Sciences, University of Patras, Patras, Greece

Margarita-Ioanna Koufaki
The Golden Helix Foundation, London, United Kingdom

Paula K. Lorgelly
School of Population Health and Department of Economics, University of Auckland, Auckland, New Zealand

Surakameth Mahasirimongkol
Ministry of Public Health, Mueang Nonthaburi, Thailand

Deborah Marshall
Department of Community Health Sciences, University of Calgary, Calgary, AB, Canada

Christopher McCabe
Queen's Management School, Queen's University, Belfast, Northern Ireland

Christina Mitropoulou
The Golden Helix Foundation, London, United Kingdom; Department of Genetics and Genomics, College of Medicine and Health Sciences, United Arab Emirates University, Al-Ain, United Arab Emirates

George P. Patrinos
Department of Pharmacy, School of Health Sciences, University of Patras, Patras, Greece; Department of Genetics and Genomics, College of Medicine and Health Sciences, United Arab Emirates University, Al-Ain, United Arab Emirates; Zayed Center of Health Sciences, United Arab Emirates University, Al-Ain, United Arab Emirates

Katherine Payne
Manchester Centre for Health Economics, Division of Population Health, Health Services Research & Primary Care, School of Health Sciences, The University of Manchester, Manchester, United Kingdom

Kathryn Philips
Department of Clinical Pharmacy, Center for Translational & Policy Research on Personalized Medicine (TRANSPERS), University of California, San Francisco, CA, United States

Samantha Pollard
Cancer Control Research, BC Cancer, Vancouver, BC, Canada

Kinasih Prayuni
Genetics Research Centre, YARSI University, Jakarta, Indonesia

Dean A. Regier
Cancer Control Research, BC Cancer, Vancouver, BC, Canada; School of Population and Public Health, University of British Columbia, Vancouver, BC, Canada

Saowalak Turongkaravee
Faculty of Pharmacy, Mahidol University, Salaya, Thailand

Athanassios Vozikis
Economics Department, University of Piraeus, Piraeus, Greece

Sukanya Wattanapokayakit
Ministry of Public Health, Mueang Nonthaburi, Thailand

Deirdre Weymann
Cancer Control Research, BC Cancer, Vancouver, BC, Canada

Sarah Wordsworth
Health Economics Research Centre, Nuffield Department of Population Health, University of Oxford, Oxford, United Kingdom; Oxford National Institute for Health Research Biomedical Research Centre, Oxford, United Kingdom

Rika Yuliwulandari
Department of Pharmacology, Faculty of Medicine, YARSI University, Jakarta, Indonesia; Genetics Research Centre, YARSI University, Jakarta, Indonesia

Hilyatuz Zahroh
Genetics Research Centre, YARSI University, Jakarta, Indonesia

Preface

We are delighted to deliver to the scientific community the textbook *Economic Evaluation in Genomic and Precision Medicine*. This is the second book on this topic published by Elsevier, following the successful launch of a similar title in 2015 by Fragoulakis, Mitropoulou, Williams, and Patrinos that was well received by the international scientific community.

Economic Evaluation in Genomic and Precision Medicine is a fundamental new discipline that aims to facilitate the integration of personalized medicine itnerventions into mainstream medical practice. This textbook provides an in-depth examination of essential concepts, protocols, and applications of economic evaluation in genomic and precision medicine. The 10 chapters in this textbook have been contributed by leading international human geneticists and health economists who are actively involved in economic assessment of genomic and precision medicine. These chapters discuss how to effectively assess the costs and outcomes of different genomic care pathways, implement cost-effective genomic interventions, and generally enhance the value of genomic and precision health care.

Foundational chapters and discipline-specific case studies cover topics ranging from economic analysis of genomic trial design to health technology assessment of next-generation sequencing, ethical aspects, economic policy in genomic medicine, and pricing and reimbursement in clinical genomics. Across these chapters, we provide multiple examples of how genomic technologies can be leveraged to simultaneously reduce costs and enhance the value of health care. As such, this textbook will allow health economists to learn to apply sound economic analysis to precision medicine, and healthcare professionals and clinical researchers will discover methods to better assess the value of genomic diagnostics and improve clinical genomic practice and drug discovery. The textbook also provides in-depth course material for professional, graduate, and undergraduate courses, examining innovative applications of economic evaluation in diverse clinical settings.

Constructive comments and feedback from colleagues and attentive readers would be gratefully received and will contribute to improving the contents of this book in future editions. We are grateful to the editors Peter Linsley and Kristi Anderson at Elsevier, who helped us in close collaboration to overcome encountered difficulties. We also express our gratitude to all contributors for delivering outstanding compilations that summarize their experience and many years of hard work in their field of research. We are indebted to Mark Rogers, who was responsible for the design and the cover of this book, and to the copy editor, who refined the final manuscript prior to its publication.

Last, but not least, we wish to cordially thank our families for their patience and continuous support, from whom we have taken a considerable amount of time to devote to this project.

Christina Mitropoulou

The Golden Helix Foundation, London, United Kingdom

Department of Genetics and Genomics, College of Medicine and Health Sciences, United Arab Emirates University, Al-Ain, United Arab Emirates

Sarah Wordsworth

Health Economics Research Centre, Nuffield Department of Population Health, University of Oxford, Oxford, United Kingdom

Oxford National Institute for Health Research Biomedical Research Centre, Oxford, United Kingdom

James Buchanan
Health Economics Research Centre, Nuffield Department of Population Health,
University of Oxford, Oxford, United Kingdom
Oxford National Institute for Health Research Biomedical Research Centre,
Oxford, United Kingdom

George P. Patrinos
Department of Pharmacy, School of Health Sciences,
University of Patras, Patras, Greece
Department of Genetics and Genomics, College of Medicine and Health Sciences,
United Arab Emirates University, Al-Ain, United Arab Emirates
Zayed Center of Health Sciences, United Arab Emirates University,
Al-Ain, United Arab Emirates

CHAPTER

1 Introduction to economic evaluation in health sciences

Christina Mitropoulou[a,b], Vasileios Fragoulakis[a], James Buchanan[c,d], Sarah Wordsworth[c,d], and George P. Patrinos[b,e,f]

[a]*The Golden Helix Foundation, London, United Kingdom,* [b]*Department of Genetics and Genomics, College of Medicine and Health Sciences, United Arab Emirates University, Al-Ain, United Arab Emirates,* [c]*Health Economics Research Centre, Nuffield Department of Population Health, University of Oxford, Oxford, United Kingdom,* [d]*Oxford National Institute for Health Research Biomedical Research Centre, Oxford, United Kingdom,* [e]*Department of Pharmacy, School of Health Sciences, University of Patras, Patras, Greece,* [f]*Zayed Center of Health Sciences, United Arab Emirates University, Al-Ain, United Arab Emirates*

1.1 Introduction

Economics is "the social science that deals with the production, distribution and consumption of goods and services." This concept refers to a set of interdependent social activities through which people produce, share, and consume products, either individually or collectively, to meet their needs, improve their living conditions, and gain prosperity [1,2].

According to this definition, economics is a social science. *Social* is the term that refers to society, its institutions, and its members, as well as the relationships between these groups. Hence, economics examines the specific economic relationships and actions of social groups and individuals as they take place in a society at a particular time. Economics is also related to politics. Politics and economics have always been substantially and historically interwoven, since there can be no economic activity without an organized system for its rules of conduct that is implemented by a central authority.

While economics deals with the production of goods, it also describes the processes and methods by which raw materials and technical expertise are transformed into useful products ready for exchange. Economics is therefore also technical in character. At the same time, the concept of "distribution" is related to the way that people acquire the legal right to own the produced goods and, to a certain extent, economics reflects the degree of social consensus regarding the equalization of wealth and societal preferences related to social justice and ethics. In addition, the concept of "consumption" involves the notion of subjective preference of one good over another; therefore economic approaches can be used to define dynamic models of human behavior in the context of economic activity.

Economics, as a discipline, describes a way of thinking about situations and problems. Economists derive economic principles, which can be used to formulate policies to solve complex economic problems. Since, as individuals and as a society, we do not have enough **goods, services, or economic resources** to do or have everything we want, we have to make **choices**. Which of our needs do we satisfy, and which ones do we leave unsatisfied? By making these choices we are rationing **our limited,**

Economic Evaluation in Genomic and Precision Medicine. https://doi.org/10.1016/B978-0-12-813382-8.00005-7

scarce resources among the infinite variety of competing goods and activities. As individuals, we have to make choices about what types and amounts of goods and services we will purchase, and how we will allocate our time. As a society, we have to make choices about who will receive what types and amounts of goods and services, and how those goods and services will be produced and exchanged.

Economics is the science of constant choices about what should be produced and by whom, how it will be produced, how it will be distributed among people, and who will consume it now and in the future.

In this book, we focus on a particular field within economics called "health economics" and specifically a subset of this field: economic evaluation of personalized healthcare. Health economics is concerned with the choices of society regarding the healthcare sector [3]. Which health services should we produce? Who will pay for them? The government or the individual? Who is entitled to health care? Those who pay taxes or the unemployed as well? Who will "consume" health services and how will consumption be prioritized? Who will "produce" health services? The government or the private sector? These are just a few of the thousands of questions that could be raised in the subject. In this book we focus specifically on the "economic evaluation of health services," which is an approach that combines methods from classical economics, mathematics, statistics, and health sciences. Economic evaluations generate evidence on the costs and benefits of new health technologies compared to existing practice, and this evidence can be used to determine which technologies should be provided by a healthcare system. For example, economic evaluations could produce evidence on whether it is socially worthwhile for a healthcare system to offer genomic testing for specific types of patients, and what the benefits of testing might be for society or for insurance companies. We provide a more detailed introduction to economic evaluation in Section 1.3. In the following section, we introduce the reader to the key concepts of economics and consider how these concepts apply within the field of health economics.

1.2 Key economic concepts and terminology used in health economics

1.2.1 Budget constraints and demand

If an economic agent (e.g., a hospital, a private patient, an insurance fund, etc.) has a "budget constraint," this means that they have a limited amount of money to invest, which restricts their consumption of goods and services. In the short term, an agent could borrow money to relax their budget constraint. However, in general, the level of production of an organization or a state defines their budget constraint, which then determines their maximum consumption. For example, wealthy countries with high incomes can invest in health infrastructure, while others with low productivity and incomes cannot invest at the same level. This is also true for private organizations and individuals; those who produce useful goods and are paid for them receive more resources and have more consumption opportunities.

Budget constraints should be expressed in real terms, that is, the number of real, natural products that can be purchased, rather than in abstract measurement units such as monetary units (e.g., Euros, dollars, etc.). In economic analyses carried out on health services, the conclusions are often different if the budget constraint changes. For example, an expensive new healthcare technology could be adopted (and reimbursed) by one resource-rich healthcare system, whereas a second healthcare system with fewer resources considers this technology to be prohibitively expensive. Economic analyses are therefore commonly reproduced in different countries and healthcare systems to reflect differences in disease management and budget constraints.

The concept of the budget constraint can be represented graphically. Let us assume that a society has only two commodities—"apples" and "oranges"—and 100 monetary units (e.g., dollars) that can be allocated between those two commodities. Let us further assume that apples cost 10 per unit and oranges cost 20 per unit. The following simple equation describes the budget constraint of this society:

$$\text{Total monetary units} = (\text{Unit Price\_of\_Orange} \times \text{Quantity of Orange}) + (\text{Unit\_Price of Apple} \times \text{Quantity of Apple})$$

Given that the prices and the available income are known, there are several feasible combinations of those two goods that could be purchased. Fig. 1.1 describes the set of options available.

This set of options takes the form of a straight line. If the income of society increases (decreases) this line shifts outwards (inwards). If the relative prices of the goods change, the slope of this line would also change accordingly. If these hypothetical goods are replaced by other, realistic commodities ("healthcare services," "education," etc.), the budget constraint shows the rate of substitution when we want to choose between these commodities.

The preceding example illustrates a common scenario: often in life a choice must be made between two alternatives that both offer benefits. The concept of "opportunity cost" captures this idea in economics. This term reflects the fact that a person must give up some of one resource in order to get some of another, and is defined as "the amount that the person has to sacrifice in order to enjoy a certain quantity of some other good (opportunity). For example, it makes no sense to mention how valuable a project or product is if we do not consider which other projects will be canceled in its favor. The economic evaluation of health services is driven by this notion: new therapies can only be appropriately evaluated if the opportunity cost of alternative uses of this money is considered.

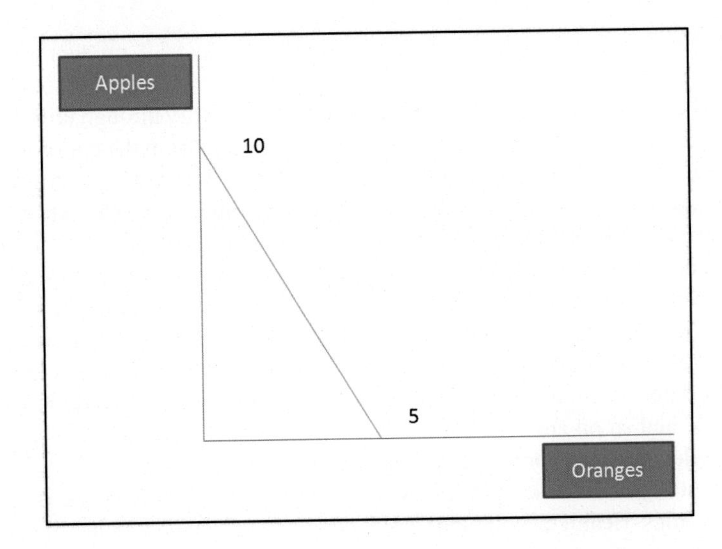

FIG. 1.1

Budget constraint.

For instance, one aim of genomic medicine is to exploit an individual's genomic profile to personalize therapeutic interventions, which has the potential to improve patient quality of life and reduce healthcare expenditure. To this end, and to maximize the outcome of the various genome-guided interventions, one needs to create an appropriate healthcare environment by identifying and adequately targeting evidence gaps for the importance of public health genomic priorities. This way not only will clinicians be persuaded for the need to implement genome-guided decision-making but also policymakers and regulators for the need to adopt and reimburse genomic tests from insurance funds and to prioritize research, development, and innovation in all genomic medicine disciplines. This process must be implemented in a way that provides an understanding of the relative benefits and drawbacks of alternative strategies (within and outside of the genomic sector) to ensure that patients receive not only effective but also economically efficient care and that healthcare systems remain sustainable in the future. It must, however, be noted that in the healthcare sector the role of "competition" amongst providers is contested and the debate about the potential role for competition is often polarized.

The concept of "demand" is intrinsically linked to the concept of the budget constraint. Demand describes the relationship between the quantity to be purchased or consumed at each given price and is determined by the available income and the preferences of consumers. Conversely, "need" describes what it is acceptable to provide in a society in which costs should not be a limiting factor. The extent to which any gap between "demand" and "need" is important is a political consideration with social implications rather than a technical problem to be resolved using economic approaches.

In conclusion, the budget constraint indicates the set of combinations that a consumer can afford with a given income. But how many apples and oranges should we buy in the end? How much money should be invested in healthcare services and education? Any combination of goods provides a different level of satisfaction or "utility," a concept that we describe in the following section.

1.2.2 Utility

The concept of "utility" is closely linked to the concept of a "product." A product is a good or service that is valued by an individual or a group and that can be given in exchange for something else that is also valuable. Consumption of a product generally increases prosperity through utility. Utility is a term that describes the feeling of joy, wellbeing, or satisfaction resulting from the consumption of a desired product. Utility is a subjective notion in that the amount of utility provided by a product can vary considerably amongst individuals or for the same individual over time. Marginal utility is simply the additional, or incremental, satisfaction derived from the consumption of the next unit of the product. The basic axiom of economics is that greater consumption provides greater prosperity, and this continues indefinitely (i.e., there is no point of saturation, although it is commonly assumed that there is diminishing marginal utility with increased consumption [4]). Better working conditions, lower unemployment, faster computers, safer mobility, more opportunities for recreational foreign travel, better health services, a socially safer environment, the ability to offer young people a modern, technical, anthropocentric education, and so on are constant needs for society and ultimate goals for individuals and societies. Fig. 1.2 illustrates the concept of utility graphically.

The concept of utility in the case of health goods is slightly different from the other goods that are considered in economics. Health technologies allow people to achieve better quality of life with fewer side effects or mobility challenges and more social activity and years of life gained and thus these technologies generate utility by diminishing negative consequences.

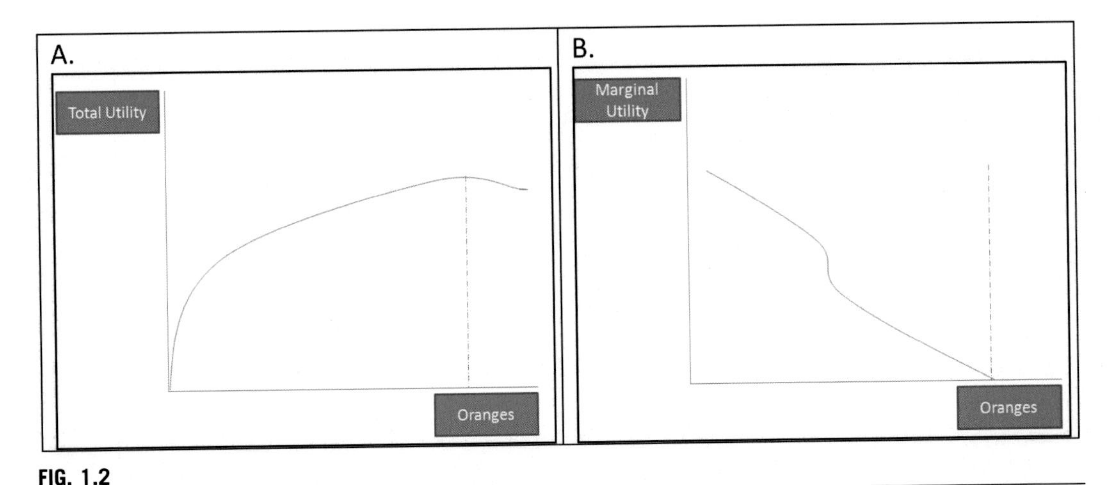

FIG. 1.2

Total (A) and marginal utility (B).

This concept of utility is frequently demonstrated by a country's economic policies. Often, a basic bundle of goods is provided as social welfare to a large proportion of citizens, as these people are in the part of the curve in Fig. 1.2 where a small surplus can greatly increase their utility. Conversely, it is more difficult to increase utility for more affluent citizens because additional prosperity requires much more consumption of goods, and a small reduction in their income does not have a strong impact on overall utility. Such attempts to redistribute utility through redistribution of income to society are carried out via taxation.

We can also extend the concept of utility from one good to two goods. Take another look at Fig. 1.2. You may have found that the consumption of extra units of "oranges" beyond a certain limit provides little extra pleasure, while the unit cost of buying/producing each extra orange is fixed. What would you do if you were given the opportunity to buy another good instead (e.g., "apples")? It would probably be rational for you to avoid consuming additional oranges and to buy apples instead. This seems reasonable, as the first oranges you would purchase would obviously not be subject to the limiting rule of decreasing marginal utility. The concept of the "indifference curve" is used to explore the relationship between the utility gained from two products. We describe this concept in the following section.

1.2.3 Indifference curves

An indifference curve describes all the combinations of two products that provide the consumer with the same level of satisfaction or, alternatively, the combinations of two products among which the consumer is indifferent. Indifference curves do not overlap; hence an individual cannot be located on more than one indifference curve simultaneously.

To illustrate this, let us assume that a consumer receives the same level of utility (satisfaction) from consuming eight apples and one orange, or from consuming one apple and nine oranges, or from consuming five apples and five oranges, and so on. Although the consumer may be indifferent to various combinations of products along a particular indifference curve, they will always prefer combinations of goods and services that place them on a higher indifference curve, that is, further towards the northeast

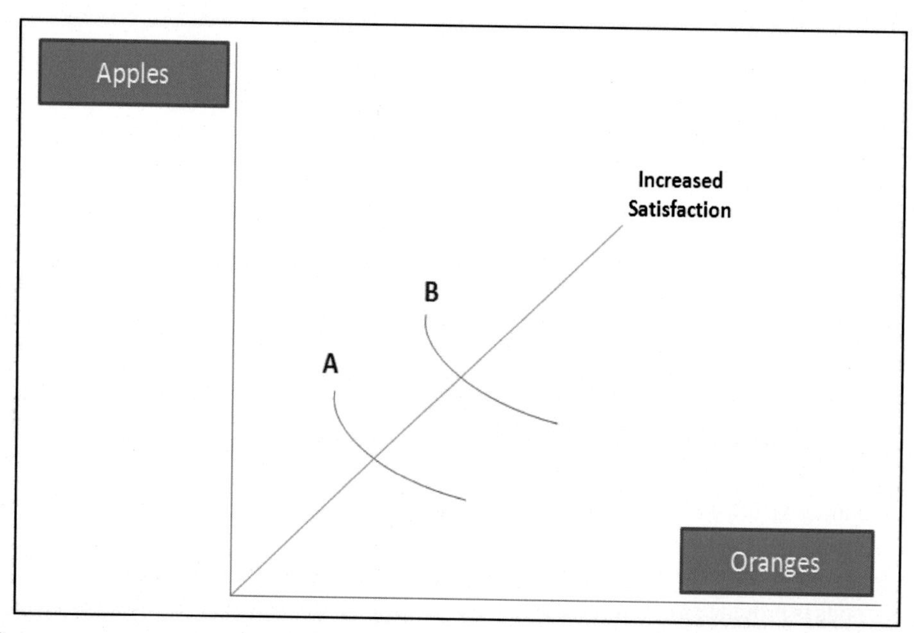

FIG. 1.3

Indifference curves.

point on the chart [5]. In Fig. 1.3., the individual is indifferent among the combinations of apples and oranges along curve A; however, any combination on curve B is preferred to any combination on curve A since curve B provides the consumer with a greater level of satisfaction.

As shown in Fig. 1.3, indifference curves are negatively sloped and have a convex shape. In this example, the consumer is initially willing to trade a relatively large quantity of oranges for an additional apple, but as they obtain more apples, they are willing to give up fewer oranges. More generally, one can view indifference curves A, B, and so on as representing levels of satisfaction that increase from "pleased" to "very pleased" to "delighted," for example. Indifference curves therefore provide information on the combinations of two products that allow a consumer to maintain a given level of satisfaction.

So far, we have only discussed combinations of products that yield the same level of satisfaction, without considering the cost of these combinations. However, it is of interest to identify combinations of two products on a given indifference curve that yield a certain level of satisfaction at a minimum cost. Graphically, this is achieved by identifying the indifference curve that is tangent to the budget constraint, as shown in Fig. 1.4.

The point at which curve B is tangent to the budget constraint (point K) identifies the combination of apples and oranges that yields the most utility for the available budget. If a consumer makes this choice, they will likely be very pleased with the quantity of apples and oranges that they receive. If the shape of the indifference curves changes over time (reflecting new preferences), or if the slope of the budget constraint changes (because relative prices or income changes), the point of optimization will change accordingly.

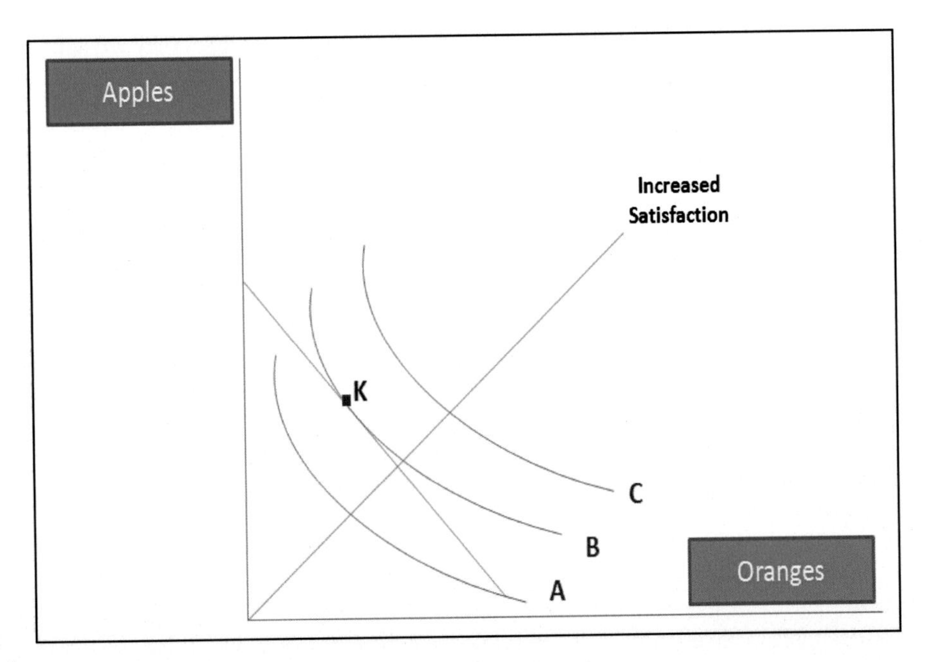

FIG. 1.4

A budget constraint plotted alongside a map of indifference curves.

From a healthcare perspective, to understand the consumption decisions of consumers we therefore need to know (1) the options available, (2) their prices, (3) consumer preferences, and (4) the level of disposable income. Even within this simple analytical framework, it is clear that there are some unique features of the health sector. For example, preferences are rarely fully defined (due to absence of information), needs vary by personal characteristics such as income, health status, and age, the options available cannot be fully evaluated by non-experts, and health professionals (specialists) must sometimes balance a societal obligation to contain costs with personal incentives to oversupply health care. Moreover, there are many competing alternative uses of the healthcare budget (should we spend money on cardiac care or primary care infrastructure?), and the health sector itself must also compete for funding with sectors such as defense, education, and energy. In practice, the healthcare policy of a country often reflects political decisions or has a historical context, and the scientific and/or analytical determination of the most rational behavior based on available data may be overlooked. To comment on such decisions, the following section introduces the economic concept of "social welfare."

1.2.4 Social welfare

It is often assumed that individual utility is related to personal consumption and is not influenced by the satisfaction of other consumers. Under this simplistic assumption, social catastrophe, the collapse of a country, the general state of poverty, and so on would not diminish the wellbeing of a consumer if their consumption pattern was unaffected.

In practice, this assumption cannot explain observed behavior, such as making charitable donations. A more rational approach is to assume that human behavior is determined by the pursuit of survival both for the individual and for the group or environment in which they are active, so as to provide a sufficient balance to enable safe social activity. Based on this assumption, no economic operator or individual will prosper if they do not personally contribute to the prosperity of society as a whole.

In Section 1.2.3 we noted that it is unclear how resources in an economy should be allocated to maximize social wellbeing. **A simple assumption is often made that social welfare is the sum of the welfare of its constituents, with each person making the same contribution to the social wellbeing**. In this scenario, an improvement in the welfare of any one person is desirable. **Another way to maximize social welfare is to place weights on the preferences of different population groups**. For example, those who actively contribute to a society could be supported, leading to resource allocation decisions being made in their favor. Each of these approaches implies that different resource allocation decisions should be made in the field of economy and health when managing a population.

If the first assumption applies, it is only important that there is an improvement in social welfare; the distribution of social welfare is unimportant. Improving the welfare of richer members of society while increasing the gap between them and the poorer members of society is not a problem in this scenario. The concept of "Pareto improvement," named after the Italian theoretician Vilfredo Pareto who formulated it, is closely linked to this idea. According to this principle, if the welfare of one person can be improved without reducing the welfare of others by economic or political action, this action is said to be desirable. This principle is a central doctrine of economics, but it has two disadvantages: (1) it is not often observed in practice and (2) it does not consider social justice. Indeed, it is likely that a decision on resource allocation in the health sector (e.g., insurance coverage of a medicine granted after genetic testing) will affect a large number of social groups and in different directions. In this example, patients who receive a drug at the correct dose after the genetic test may improve their condition but at the expense of other patients who may be burdened with additional tax payments to cover the expenses of the patients who benefit. Full economic analyses should be undertaken to estimate the additional costs and benefits of the new treatment, and to consider whether the benefits justify the costs. In this case, the concept of Pareto improvement is not particularly helpful.

Alternatively, the Kaldor-Hicks criterion proposes that one should calculate whether the benefit to one social group overcomes the burden on another social group to determine pure societal profit. The Kaldor-Hicks criterion requires that it is possible for those who have benefited from a policy to compensate those who have been harmed (in order to bring them back to their former state), yielding a net social benefit. This approach is potentially more attractive, but it is still difficult to determine the benefits arising to different groups and to find an objective way of achieving this compensation in practice. The healthcare sector is characterized by the presence of multiple interested parties, each trying to achieve multiple goals. For example, the state (the government) may be interested in reducing spending and abiding by a budget constraint while at the same time providing a minimum level of health care. Healthcare providers want to maximize profits and increase their market share by increasing the demand for health care via health marketing. Conversely, patients want access to innovation at the lowest possible cost and the hope of a better and longer life. It is difficult to have clear rules regarding the maximization of social welfare in such a scenario, and often the government must negotiate with healthcare providers, for example, setting sales limits or maximum prices. The broad aim of such measures is to ensure that all stakeholders can operate in a stable environment with concrete, sustainable conditions.

With respect to the third criterion mentioned—fixing welfare at the level of the least prosperous social group—it should be noted that access to health care is a socially sensitive issue, and almost all governments aim to provide a "safety net" that will provide, among other things, access for the least prosperous members of society to the public healthcare system. Often, social considerations prevail over economic profitability, with a view to strengthening social cohesion. A further consideration is that the introduction of a new health technology can have different effects on different people depending on the decisions taken. For example, increasing life expectancy in an elderly person may be valued differently compared to a similar change in life expectancy in a younger patient.

1.3 Economic evaluation

So far in this chapter we have introduced economics as a social science, described some key concepts, and considered how these concepts apply within the field of health economics. In this section, we build on this by explaining how we can use economic evaluation to generate evidence on the costs and benefits of new health technologies compared to existing practice. We also consider how this evidence can be used to determine which technologies should be provided by a healthcare system.

1.3.1 Rationale for economic evaluation

In Section 1.1 we introduced the idea that individuals and societies must choose how to allocate scarce resources among competing alternative uses. Allocating scarce resources efficiently ensures that welfare gains to society are maximized. In a free market, we can be relatively confident that resources will be allocated efficiently because consumers and producers are free to interact, with the price level used to equalize demand and supply. However, the existence of a free market requires several assumptions to hold. These include the existence of perfect knowledge and certainty, no externalities (consequences of economic activity that are experienced by unrelated third parties [6]), and the presence of many producers with no market power. These requirements are met relatively easily in some sectors of the economy, for example, markets for consumer goods such as food. However, in other sectors such as health care it is not as straightforward, and alternative approaches to allocating scarce resources are required.

Why is resource allocation a particular challenge in health care? First, the free market fails because individuals cannot reliably predict when they will require health care. This creates a need for an insurance market. However, healthcare insurance markets face two key challenges: moral hazard (if an individual possesses insurance, they may change their behavior) and adverse selection (an individual often knows more about their risk status than those who are offering insurance). Second, externalities exist in the healthcare sector. For example, if you are vaccinated, this has a personal benefit, and is also beneficial for the rest of society. Third, clinicians act on behalf of patients to both demand and supply health care. The relationship between clinicians and patients, often called an imperfect principal agent relationship, could lead to more health care being consumed than is optimal. Finally, as mistakes in health care can have serious consequences, healthcare practitioners must be professionally licensed. However, this "market power" can provide practitioners with an incentive to reduce the supply of health care. These challenges are not unique to health care; externalities are also a consideration when making resource allocation decisions related to environmental interventions, for example. However, when combined, these problems can prevent the free market from ensuring the efficient allocation of resources in health care. Given this, government intervention is commonly required to allocate healthcare resources efficiently [6].

The extent to which governments intervene to allocate resources in health care varies across countries. For example, in the United Kingdom, the core principles of the National Health Service (NHS) include commitments that health care will be available to all and free at the point of delivery [7]. Given this, it has been suggested that the NHS compromises on the range and quality of services offered to patients [8]. Consequently, the most clinically effective interventions are not always offered to patients or are only offered to specific subgroups. Resource allocation decisions such as these can be made in a variety of ways (e.g., using waiting lists), but these approaches can lead to resources being allocated in a non-systematic or subjective manner. Alternatively, these decisions can be made using a decision tool in a systematic and explicit way. One such tool is economic evaluation.

1.3.2 Types of economic evaluation

An economic evaluation is a "comparative analysis of alternative courses of action in terms of both their costs and consequences" [3]. By computing and comparing the costs and outcomes of different healthcare interventions, economic evaluation can be used to identify interventions that represent value for money. Economic evaluations of healthcare interventions can either take a "welfarist" or "extra-welfarist" approach. If a welfarist approach is used, the broad aim of the economic evaluation is to maximize societal welfare with reference to a societal budget constraint [9,10]. If an extra-welfarist approach is used, the broad aim is to maximize health effects in a resource-constrained health system [10–12]. The key difference is that the extra-welfarist approach assumes that the budget is fixed, and available resources need to be efficiently allocated (e.g., should more of the health service budget be spent on genomic testing?), whereas the welfarist approach can address whether budgets have been efficiently determined in the first place (e.g., should we spend more on the health service or education?). When operationalized, both approaches measure healthcare costs in monetary terms but differ in the measurement of health outcomes.

1.3.2.1 The extra-welfarist approach to economic evaluation

The extra-welfarist approach can be operationalized in several ways. Cost-minimization analyses, in which only the costs of interventions are compared because their health outcomes are the same, were once commonly undertaken, but this approach is no longer considered to be appropriate by health technology assessment agencies (one exception is economic evaluations of biosimilars) [13–16].

Instead, the extra-welfarist approach is usually operationalized via cost-effectiveness analysis (CEA) or cost-utility analysis (CUA). Both approaches measure outcomes in health-related terms. In CEAs, natural units are used such as cases detected or number of life years gained from using an intervention. In CUAs, quality-adjusted life years (QALYs) are used, which combine information on longevity and quality of life (disability-adjusted life years are also used in some settings). In both cases, the cost and outcome differences between a new intervention and its comparator are calculated. If a new intervention is both less costly and more effective than its comparator, the new intervention dominates, whereas if the new intervention is more costly and less effective, the comparator dominates. When a new intervention is both more expensive and more effective, the cost and outcome differences are expressed as an incremental cost-effectiveness ratio (ICER). Fig. 1.5 summarizes these different scenarios and presents the cost-effectiveness plane.

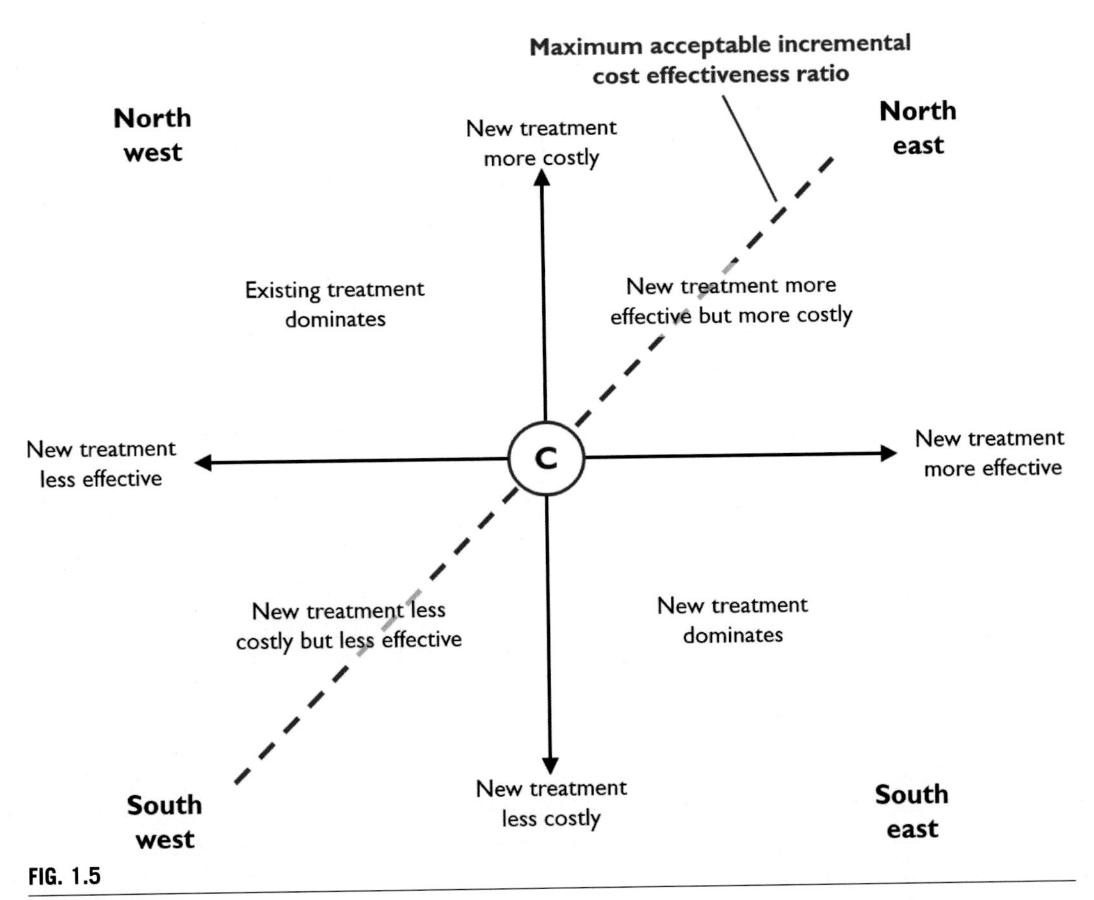

FIG. 1.5

The cost-effectiveness plane.

If an ICER has been calculated, this is usually compared to a threshold ICER that represents the maximum amount society is willing to pay for an additional unit of health outcome. If the ICER falls below this threshold, the new intervention is said to be cost-effective. Threshold ICERs vary between countries; in England, a threshold of £20,000–£30,000 is commonly applied, while in the United States, a threshold of $50,000 is often used.

1.3.2.2 The welfarist approach to economic evaluation

The welfarist approach is usually operationalized using cost-benefit analysis (CBA). In CBA, health outcomes are valued in monetary terms using one of several potential analytical approaches [3]. For example, the human capital approach uses market wage rates to value healthy time, which allows the impact of an intervention on future earnings to be calculated. Alternatively, the revealed preference approach is an indirect method that considers the interaction between a non-marketed good and a market for another good (e.g., the travel costs incurred to use a particular health facility). Whichever approach is used, results are expressed as either a ratio of benefits to costs or as a sum representing the net benefit of one intervention compared to another. In this scenario, if a new intervention generates a positive net benefit (or a ratio of benefits to costs greater than one), it should be adopted. A key advantage of this approach is the use of a common unit (money) to value outcomes. This allows the net benefit of healthcare interventions in a variety of disease areas to be easily compared.

1.3.2.3 Welfarism versus extra-welfarism: Which approach should be used?

There are clearly fundamental differences in the aims and practical application of the welfarist and extra-welfarist approaches to economic evaluation. Given this, which approach should be used to allocate healthcare resources? The most commonly applied approach has varied over the past 60 years. The welfarist approach initially dominated health economics during the 1960s and 1970s, until several limitations were identified. For example, CBA may favor diseases of the affluent over those of the poor [9,11,17], and the human capital approach may discriminate against those outside the labor force [18–20]. The extra-welfarist approach consequently became more popular in the 1980s, which coincided with the development of key metrics such as the QALY [21] and an increased demand for economic evidence. This approach was subsequently adopted by most HTA agencies around the world. In the past 20 years, some dissatisfaction has arisen with the narrow health-oriented goals of extra-welfarism, particularly in contexts such as genomic testing and complex interventions, such as interventions to control the behavior of at-risk individuals (e.g., drug users) [22,23]. Practical challenges have also been identified, including a reliance on the use of rules of thumb in decision-making [24,25]. Consequently, there is increasing interest in relatively new approaches to monetize health benefits within a welfarist framework, such as the use of discrete choice experiments [26,27].

Achieving a consensus on the most appropriate approach to economic evaluation is important because it has been frequently argued that the two approaches may give different answers to the same question [19]. This is not just a theoretical concern; a recent study reported that for every four studies applying both the welfarist and extra-welfarist approaches, one shows limited or no concordance between the results of the different forms of economic evaluation [23]. This is therefore a key methodological issue in a genomic context, where a broader analytical framework might be appropriate. We will return to this debate in Chapter 3, which considers the challenges associated with undertaking an economic evaluation of a genomic test.

1.4 Concluding remarks

This chapter introduced the social science that is **economics**, and the field that will be the primary focus of this book: **health economics**. We described key concepts such as budget constraints, demand, utility, opportunity costs, and social welfare, and provided an overview of an important resource allocation tool: **economic evaluation**. Together, these ideas provide the foundation for the remaining chapters in this book. These chapters proceed as follows. Chapters 2 and 3 take a more detailed look at the two key themes of this book: personalized medicine and performing economic evaluations in a genomic context. We then present five chapters that summarize the latest economic evaluation evidence for genomics in the context of oncology (Chapter 3), rare diseases (Chapter 4), other medical disciplines (Chapter 5), and the developing world (Chapter 6). Chapters 7–9 consider technical issues related to economic modeling in this context, the use of big data to inform economic analyses, and how the results of economic evaluations can be used to guide resource allocation decisions related to precision medicine. Finally, Chapter 10 broadens our discussion to consider policy and stakeholders in economic evaluation in genomics, and issues related to pricing and reimbursement.

References

[1] Fragoulakis V, Mitropoulou C, Williams MS, Patrinos GP. Economic evaluation in genomic medicine. Burlington, CA: Elsevier/Academic Press; 2015. ISBN 978-0128014974.

[2] Samuelson P. Economics, an introductory analysis. New York: McGraw-Hill Company; 1948.

[3] Drummond MF, Sculpher MJ, Torrance GW, O'Brien BJ, Stoddart GL. Methods for the economic evaluation of health care programme. 3rd ed. Oxford: Oxford University Press; 2005.

[4] Mold JW, Hamm RM, McCarthy LH. The law of diminishing returns in clinical medicine: how much risk reduction is enough? J Am Board Fam Med 2010;23(3):371–5.

[5] Morris S, Devlin N, Spencer A, Parkin D. Economic analysis in healthcare. Hoboken, N.J.: Wiley; 2013.

[6] Donaldson C, Gerard K. Economics of health care financing: the visible hand. Palgrave Macmillan; 2005.

[7] NHS. Principles and values that guide the NHS., 2015, www.nhs.uk.

[8] Weale A. Rationing health care. Br Med J 1998;316:410.

[9] Bala MV, Zarkin GA, Mauskopf JA. Conditions for the near equivalence of cost-effectiveness and cost-benefit analyses. Value Health 2002;5:338–46.

[10] Grosse SD, Wordsworth S, Payne K. Economic methods for valuing the outcomes of genetic testing: beyond cost-effectiveness analysis. Genet Med 2008;10:648–54.

[11] Gyrd-Hansen D. Willingness to pay for a QALY: theoretical and methodological issues. PharmacoEconomics 2005;23:423–32.

[12] Johannesson M. Theory and methods of economic evaluation of health care. In: Johannesson M, editor. Developments in health economics and public policy, vol. 4. Dordrecht; Boston and London: Kluwer Academic; 1996.

[13] Dakin H, Wordsworth S. Cost-minimisation analysis versus cost-effectiveness analysis, revisited. Health Econ 2013;22:22–34.

[14] Marshall JD, Harries M, Hill D, Hill CA. PHP139—trends in the use of cost-minimization analysis in Economic Assessments submitted to the SMC. Value Health 2015;18:A94.

[15] Simoens S. Biosimilar medicines and cost-effectiveness. Clinicoecon Outcomes Res 2011;3:29–36.

[16] Briggs AH, O'Brien BJ. The death of cost-minimization analysis? Health Econ 2001;10:179–84.

[17] Donaldson C, Birch S, Gafni A. The distribution problem in economic evaluation: income and the valuation of costs and consequences of health care programmes. Health Econ 2002;11:55–70.

[18] Blumenschein K, Johannesson M. Economic evaluation in healthcare. A brief history and future directions. PharmacoEconomics 1996;10:114–22.

[19] Johannesson M. The relationship between cost-effectiveness analysis and cost-benefit-analysis. Soc Sci Med 1995;41:483–9.

[20] Johannesson M, Jonsson B. Economic evaluation in health care: is there a role for cost-benefit analysis? Health Policy 1991;17:1–23.

[21] Williams A. Economics of coronary artery bypass grafting. Br Med J (Clin Res Ed) 1985;291:326–9.

[22] Payne K, McAllister M, Davies LM. Valuing the economic benefits of complex interventions: when maximising health is not sufficient. Health Econ 2013;22:258–71.

[23] Buchanan J, Wordsworth S. Welfarism versus extra-welfarism: can the choice of economic evaluation approach impact on the adoption decisions recommended by economic evaluation studies? PharmacoEconomics 2015;33(6):571–9. https://doi.org/10.1007/s40273-015-0261-3. Review. PubMed PMID: 25680402.

[24] Tsuchiya A, Williams A. Welfare economic and economic evaluation. In: Drummond M, McGuire A, editors. Economic evaluation in health care: merging theory with practice. Oxford University Press; 2001.
[25] Brouwer WB, Koopmanschap MA. On the economic foundations of CEA. Ladies and gentlemen, take your positions! J Health Econ 2000;19:439–59.
[26] Bryan S, Dolan P. Discrete choice experiments in health economics. For better or for worse? Eur J Health Econ 2004;5:199–202.
[27] Mooney G. Beyond health outcomes: the benefits of health care. Health Care Anal 1998;6:99–105.

CHAPTER

Foundations of pharmacogenomics and personalized medicine

2

Maria Koromina[a] **and George P. Patrinos**[a,b,c]

[a]*Department of Pharmacy, School of Health Sciences, University of Patras, Patras, Greece,* [b]*Department of Genetics and Genomics, College of Medicine and Health Sciences, United Arab Emirates University, Al-Ain, United Arab Emirates,* [c]*Zayed Center of Health Sciences, United Arab Emirates University, Al-Ain, United Arab Emirates*

2.1 Introduction

Recent technological and scientific advances such as molecular biology, genetics, and genomics contribute to a better understanding of the mechanism actions of biological molecules, including xenobiotics and drugs. Nowadays, we are aware of the genetic variability among individuals and its crucial role in drug response and efficacy [1]. More precisely, genetic variability is estimated roughly to be 3.5 to 5 million genomic variants per genome, of which a significant number are rare variants that may result in observable phenotypes [2]. Variants observed within genes encoding various enzymes are of particular interest since they may affect metabolic pathways related to the absorption, distribution, metabolism, excretion, and toxicity of drug treatments. These genes are referred to as ADMET genes owing to their implication in the aforementioned procedures.

An example that highlights the clinical significance of pharmacogenomics is that approximately 25%–60% respond to their medications. This fact indicates that a massive proportion of patients do not receive the appropriate medication or that this proportion of patients is experiencing adverse drug reactions due to changing medications until a clinical benefit is observed [1].

By using the term "pharmacogenomics," we refer to "the delivery of the right drug to the right patient at the right dose." However, there are some issues that should be carefully considered before implementing pharmacogenomics into clinical practice. Amongst these issues are the following: (1) the disease heterogeneity and complexity (i.e., several phenotypes observed in a single disease) or the genetic complexity (i.e., several genes contributing to a single phenotype), (2) the small number of clinically actionable pharmacogenomic biomarkers, and (3) a plethora of public health genomics-related issues such as bioethical, societal, and economic issues [3].

This chapter provides a summary of some up-to-date examples of pharmacogenomics and personalized medicine applications in different diseases. We also shed some light on social, financial, and ethical factors that affect the implementation of pharmacogenomics in clinical practice. These factors are of great importance not only for encouragement of pharmacogenomics implementation into clinical practice but also for policy makers who are responsible for the reimbursement of pharmacogenomics testing.

Economic Evaluation in Genomic and Precision Medicine. https://doi.org/10.1016/B978-0-12-813382-8.00006-9

2.2 Implementing pharmacogenomics into clinical care

As stated, pharmacogenomics has already been applied in several medical specialties with success, thus leading to the approval of implementation of pharmacogenomics tests by the US Food and Drug Administration (FDA; www.fda.gov) and the European Medicines Agency (EMA; www.ema.europa.eu). To further encourage the implementation of pharmacogenomics in drug-development pipelines, many FDA-approved drug labels strongly recommend companion pharmacogenomics testing prior to drug prescription. Given the importance of pharmacogenomics in clinical practice, prospective application of pharmacogenomics in pharmaceutical drug development pipelines could focus on developing new molecules while reducing the associated risks and costs [1].

In the following sections, we briefly summarize some key examples of the most promising pharmacogenomics applications in clinical practice.

2.2.1 Pharmacogenomics for cancer therapeutics

Tamoxifen is a "pro-drug" and a selective estrogen receptor modulator used for prevention of breast cancer in women and treatment of breast cancer in women and men [4]. Interestingly, there is a series of studies proposing potential associations of *CYP2D6*4* and *CYP2D6*10* alleles with the metabolic activation of tamoxifen to endoxifen. For example, Borges et al. [5] showed that individuals carrying at least one null (*4) or a reduced function allele (*10) had similar endoxifen concentrations, while individuals carrying multiple copies of the gene were characterized by an increased endoxifen plasma concentration. In another study, Goetz et al. [6] concluded that women carrying the *CYP2D6*4/*4* genotype are characterized by an increased risk of disease relapse and a lower incidence of hot flashes, thus supporting further their previous observation that *CYP2D6* is responsible for the metabolic activation of tamoxifen to endoxifen.

These observations were further supported by Schroth et al. [7], who showed that women carrying the *CYP2D6*4* allele were characterized by worse treatment outcomes compared to the functional allele carriers. Taking into consideration the findings with respect to tamoxifen dosing, the FDA advisory committee updated the drug label of tamoxifen with additional clinical recommendations for its use in clinic [8].

Another example of pharmacogenomics for cancer therapeutics is the case of irinotecan, which is a camptothecin analog used for the treatment of metastatic colorectal cancer. According to findings from a previous study, the level of irinotecan metabolism was inversely associated with the number of TA repeats in the *UGT1A1* promoter region [9]. Moreover, Ando et al. [10] estimated the correlation between the *UGT1A1*28* allele and the risk of severe adverse drug reactions (i.e., diarrhea and/or neutropenia) in patients under irinotecan treatment [10]. Other studies have provided evidence for an association of the *UGT1A1*28* allele either with SN-38 glucuronidation rate [11] or with grade 4 neutropenia [12,13]. Consequently, the FDA amended the irinotecan label to include the *UGT1A1*28* allele as a risk factor for cancer patients to develop neutropenia upon irinotecan treatment.

Another compound, 5-fluorouracil (5-FU), is also known for its use in cancer treatment as a chemotherapeutic agent. The catabolism of 5-FU is mediated by the dihydropyrimidine dehydrogenase (DPD) enzyme, which is encoded by the *DPYD* gene, in approximately 80% of the administered dose [14]. Recent studies have shown that genetic variants within *DPYD* may result in DPD enzyme deficiency in approximately 4%–5% of the population. Consequently, the half-life of the drug is increased, thus leading to drug bioaccumulation and severe fluoropyrimidine-related toxicity in patients [15]. As such, [16] performed preemptive genotyping on patients treated with 5-FU for the *DPYD*2A* biomarker and then adjusted the dose accordingly. This way, it was shown that genotyping prior to 5-FU treatment can increase patient safety.

Also, meta-analysis studies have highlighted significant associations of *DPYD* variants with 5-FU-induced toxicity [17–19]. Reduced or no enzyme activity has been associated with several single-nucleotide polymorphisms (SNPs) within *DPYD*, such as *DPYD*2A*, *DPYD*13*, c.2846A>T, and c.1236G>A/haplotype B3 [20]. Of these SNPs, *DPYD*2A* has been characterized as the most common mutation associated with fluoropyrimidine-associated toxicity [21]. Although both the FDA and EMA contraindicate the use of 5-FU in patients with a known DPD deficiency, a recommendation for pre-emptive *DPYD* genotyping in cancer patients before receiving fluoropyrimidine treatment is yet to be proposed.

2.2.2 Pharmacogenomics for drug treatment of cardiovascular diseases

When combined with aspirin, clopidogrel is widely used to prevent atherothrombotic events and cardiac stent thrombosis [22]. Although clopidogrel was initially approved as an antiplatelet drug, a sub-therapeutic response was observed in almost one-fourth of individuals treated with clopidogrel. The effectiveness of clopidogrel depends on its conversion to an active metabolite by CY2C19 [23]. Hulot et al. [23] found an association between low concentration levels of clopidogrel's active metabolite and the *CYP2C19*2* allele, which is a common loss-of-function variant. Other cohort studies showed that *CYP2C19*2* and other *CYP2C19* loss-of-function alleles have been associated with non-responsiveness and adverse clinical outcomes of clopidogrel [24].

Undoubtedly, determining a patient's response to clopidogrel treatment can be a multifactorial issue that involves many genetic variants and non-genetic factors [25,26]. As such, personalized care in precision cardiovascular medicine will need to integrate "multi-omics" information with computational approaches to improve the prognosis, diagnosis, and treatment of cardiovascular disorders [27].

Another example of pharmacogenomics implementation in cardiovascular diseases is the case of coumarinic oral anticoagulants, such as warfarin, acenocoumarol, and phenprocoumon. Genetic variants in the *CYP2C9* and *VKORC1* genes could potentially result in a great inter-individual variability regarding the coumarinic drug response [28,29]. More precisely, *CYP2C9*2* and **3* alleles affect coumarin pharmacokinetics by reducing the enzymatic activity of *CYP2C9*, while a specific variant within *VKORC1* alters the pharmacodynamic response to coumarins [30]. Moreover, a genome-wide association study (GWAS) highlighted the significance of *VKORC1*, *CYP2C9*, and *CYP4F2* genes for inter-individual drug variability [29], while Pereuz-Andreu et al. suggested preemptive genotyping for *VKORC1*, *CYP4F2* genes and *CYP2C9*2* and *CYP2C9*3* alleles could assist in identifying outlier patients treated with acenocoumarol [31].

Statins are amongst the most widely prescribed drugs for the treatment of cardiovascular disorders. Findings from pharmacogenomics studies have shown that genetic variants within *SLCO1B1* affect the delivery of statins in the liver, thus leading to a reduction of statins' blood concentration [32]. Voora et al. [33] showed that the *SLCO1B1*5* allele (p.V174A, c.521T>C) is associated with an increased risk for statin-induced adverse effects (i.e., myopathy) in patients receiving statin treatment [33]. Except for *SLC01B1*, research findings suggest the involvement of other genes such as *CETP*, *KIF6*, and *CYP3A* in the metabolism of statins [34–36].

2.2.3 Pharmacogenomics for psychiatric diseases

Despite the many studies attempting to associate genetic variants with psychiatric drug response and toxicity, there are still difficulties in implementing pharmacogenomics for psychiatric diseases. Amongst these difficulties are the large inter-individual variation in drug response and the severe adverse effects experienced by patients. However, there are a few examples where pharmacogenomics has been implemented in psychiatric disease. We highlight these in the paragraphs that follow.

To begin with, lithium chloride is one of the most well-known mood stabilizers with anti-suicidal effects, mainly used as maintenance treatment in bipolar disorder (BD) [37,38]. There are some notable examples of genetic variants associated with lithium response as identified from candidate gene studies. Amongst these examples are rs1800532 within *TPH1* and rs4532 within *DRD1*, which were associated with poor lithium response [39]. Amare et al. [40] provided a review summarizing a couple of genetic variants that have been associated with lithium response, including rs3730353 within *FYN* (association with prophylactic response to lithium in BD patients); c.C973A variant within *INPP1* (association with lithium efficacy); and rs206472 within *INPP1* (association with lithium efficacy; [40]). Song et al. [41] highlighted the association of variants within *PLET1* and the response to lithium treatment. Regarding *CREB1*, two variants (rs6740584 and rs2551710) have been associated with response to lithium treatment, although these findings need to be further replicated [39]. In another study, Squassina et al. [42] identified a strong association signal between an SNP located in intron 1 of the *ACCN1* and the response to lithium in a subgroup of BD patients from Sardinia.

Notably, some of the genes already mentioned to be associated with lithium response have overlapping effects in response to antidepressants in major depressive disorder and lithium treatment response in BD, such as *SLC6A4* genomic variants [40]. Although there are many studies linking common SNPs with response to lithium, robust evidence indicating genetic variants as FDA-approved biomarkers for lithium response is yet to be found.

Although the literature indicates many genetic variants significantly associated with antipsychotic treatment response (i.e., risperidone, clozapine, quetiapine, chlorpromazine), most of these associations usually fail to be further replicated. Amongst the genes carrying variants that have been associated with antipsychotic treatment are the following: *CYP2D6*, *CYP2C19*, *COMT*, *ABCB1*, *DRD3*, *HTR2C*, *TNIK*, *RELN*, *NOTCH4*, and *SLC6A2* [43]; *HSPG2*, *CNR1*, *DPP6*, *SLC18A2*, *HLA*, and *MC4R* [44]; and *TSPO* [45].

There are numerous candidate gene studies that have been performed to identify genetic variants associated with the response to antipsychotic treatment. In these studies, researchers usually assess either pharmacokinetic- or pharmacodynamic-associated genes, or candidate genes such as dopaminergic, serotoninergic, or glutamatergic genes [45–47]. As an alternative to candidate gene studies, GWAS (as a hypothesis-free approach) have also been performed with the same goal of associating genes with antipsychotic treatment efficacy. A characteristic example is that from Yu et al.'s [48] genome-wide study, in which five novel loci were significantly associated with treatment response in a population of Han Chinese ancestry (i.e., rs72790443 in *MEGF10*, rs1471786 in *SLC1A1*, rs9291547 in *PCDH7*, rs12711680 in *CNTNAP5*, and rs6444970 in *TNIK*). Yu et al. [48] also identified three additional loci that were associated with drug-specific treatment responses (rs2239063 in *CANCA1C* for olanzapine, rs16921385 in *SLC1A1* for risperidone, and rs17022006 in *CNTN4* for aripiprazole). Although these findings need to be further replicated, this study added to our understanding of the underlying mechanisms of antipsychotic action.

Since findings usually fail to be replicated, the clinical utility for pharmacogenomics biomarkers for antipsychotic treatment mainly lies within *CYP2D6* and *CYP2C19* genes. Whole genome or whole exome sequencing (WGS or WES) studies could be an alternative solution that will make more robust the identification of genetic variants that are associated with the response to antipsychotic treatment.

2.3 Electronic tools for translating genomic findings into a clinically meaningful format

In the era of big "-omics" data, extraction of biological knowledge from the multi-omics datasets can be quite challenging. Therefore, in order to translate the (pharmaco) genomic findings into a clinically meaningful output, information extracted from already established pharmacogenomics resources (e.g., PharmGKB, CPIC, DruGeVar, etc.) needs to be combined with information from the literature as well as other genomic databases (e.g., NCBI, dbSNP, dbGAP, ClinVar, etc.). To this end, a database-driven approach should incorporate (1) literature mining/natural language processing (NLP) for extraction of putative disease-drug-gene/variant-phenotype associations from pharmacogenomics and other genomics resources/databases, (2) published literature, as well as (3) a virtual population pharmacokinetic simulator for testing putative variant-phenotype associations and assessing relevant genotype-to-phenotype covariance statistics in virtual populations. Moreover, a collaborative environment between researchers is crucial for the formation, validation, and evidential assessment of such identified associations.

Genomic databases can be classified into different categories based on the output information following a variant query. These categories are population databases, disease databases, locus/disease/ethnic/other specific databases, sequence, and pharmacogenomics databases. We summarize the most frequently used genomic databases used for variant curation in the paragraphs that follow.

Online Mendelian Inheritance in Man (OMIM; www.omim.org) is a regularly updated database primarily used for the identification of e gene-phenotype relationships and the association of genes or genetic variants with diseases [49]. More precisely, OMIM associates the genetic variants of interest with relevant information about clinical phenotypes. HbVar is a disease-specific database primarily used for characterization of hemoglobin variants and thalassemia mutations [50–52]. Furthermore, the Human Genome Variation Society (HGVS; www.hgvs.org), which evolved from the Human Genome Organization-Mutation Database Initiative (HUGO-MDI), aimed to boost the discovery and characterization of genetic variants, including the assessment of population distribution as well as the identification of any phenotypic associations. The DECIPHER database also aims to assess, share, and compare phenotypic and genotypic data by retrieving information from multiple databases, thus leading to interpretation of the clinical relevance of the variants of interest (Table 2.1).

Another computational-driven approach is the development and use of genome browsers. Amongst the most frequently used genome browsers are the University of California, Santa Cruz (UCSC) Genome Browser, EBI's Ensembl, and the Genome Data Viewer [68–70]. These browsers provide genomic context for individual features, such as genes or disease loci, in a viewable and extractable output format and they can be also used for variant analysis and curation.

Nevertheless, it is important to highlight that not all genomic databases and browsers meet the quality requirements for variant curation and interpretation. Amongst the issues that need to be

Table 2.1 Summary of some of the most widely used genomics and pharmacogenomics databases, including: name of database, database type, database description, database reference, and database URL.

Database type	Database	Description	Reference(s)	Link
Population databases	Exome Aggregation Consortium (ExAC)	Browser with gene- and transcript-centric displays of variation	[53]	http://exac.broadinstitute.org
	Genome Aggregation Database (gnomAD)	Browser similar to ExAC; it supports aggregation of genome data	[54]	https://gnomad.broadinstitute.org
	Exome Variant Server	Browser containing exome sequence data on 6503 individuals; it displays MAFs for African Americans and European Americans	Exome Variant Server, NHLBI GO Exome Sequencing Project (ESP), Seattle, WA (URL: http://evs.gs.washington.edu/EVS/) (09/2019 accessed)	https://evs.gs.washington.edu/EVS/
	dbSNP	Free public archive for genetic variation (multi-species)	[55]	https://www.ncbi.nlm.nih.gov/snp/
	dbVar	NCBI's database of human genomic structural variation	–	https://www.ncbi.nlm.nih.gov/dbvar/
	FINDbase	A database for clinically relevant genomic variant allele frequencies for various populations worldwide	[56]	http://www.findbase.org
Disease databases	ClinVar	Free public archive of associations among sequence variation and human phenotypes	[57]	https://www.ncbi.nlm.nih.gov/clinvar/
	dbGAP	NCBI's archive of data and results from studies investigating the interaction of genotype and phenotype in humans	–	https://www.ncbi.nlm.nih.gov/gap/
	OMIM	Regularly updated catalog of human genes, genetic disorders, and traits, focusing on the gene-phenotype relationship	[58]	https://www.omim.org

Table 2.1 Summary of some of the most widely used genomics and pharmacogenomics databases, including: name of database, database type, database description, database reference, and database URL—cont'd

Database type	Database	Description	Reference(s)	Link
Locus/disease/ethnic/other-specific databases	Human Gene Mutation Database	Database of published germline mutations that are closely associated with human inherited disease	[59]	http://www.hgmd.cf.ac.uk/ac/index.php
	Human Genome Variation Database	Curated catalog of genome variants published in the peer-reviewed *Data Reports* and relevant articles in *Human Genome Variation*	–	https://hgv.figshare.com
	Human Genome Variation Society	Catalog containing genomic variant information and associated clinical variations	[60]	https://www.hgvs.org
	Leiden Open Variation Database	Locus-specific database containing information about identified genetic variants and associations with disease phenotypes	[61]	https://www.lovd.nl
	DECIPHER	Database of genomic variation data from analysis of patient DNA; mapping the clinical genome	[62]	https://decipher.sanger.ac.uk
	HbVar	Database with information on the genomic alterations leading to hemoglobin variants and hemoglobinopathies	[51]	http://globin.cse.psu.edu/globin/hbvar/
Sequence databases	NCBI Genome	Database with information on genomes, including sequences, maps, chromosomes, assemblies, and annotations	–	https://www.ncbi.nlm.nih.gov/genome
	RefSeqGene	Database that defines genomic sequences to be used as reference standards for well-characterized genes	[63]	https://www.ncbi.nlm.nih.gov/refseq/rsg/

Continued

Table 2.1 Summary of some of the most widely used genomics and pharmacogenomics databases, including: name of database, database type, database description, database reference, and database URL—cont'd

Database type	Database	Description	Reference(s)	Link
	Locus Reference Genomic	Curated database containing genomic, transcript, and protein reference sequences for reporting clinically relevant sequence variants	[64]	https://www.lrg-sequence.org
	MitoMap	A compendium of polymorphisms and mutations in human mitochondrial DNA	[65]	https://www.mitomap.org
Pharmacogenomics databases	PharmGKB	Database containing information on the impact of human genetic variation on drug response	[66]	https://www.pharmgkb.org
	CPIC	Database containing information on gene/drug clinical practice guidelines	[67]	https://cpicpgx.org

addressed is the privacy issue, which could lead to the retraction of datasets from browsers, as well as the need for standardization and proper documentation towards a more effective use of the genome browsers [71].

2.4 Assessing the ethical, societal, and financial aspects of personalized medicine

Undoubtedly, the application of next-generation sequencing in pharmacogenomics, either as pharmacogene resequencing in low-resource settings or as whole exome or whole genome sequencing focusing only on the variants called within the pharmacogenes, has led to creation of more robust and efficient clinical pharmacogenomics pipelines for the identification of biomarkers [72].

However, there are still challenges and difficulties that need to be overcome prior to implementing pharmacogenomics in clinical practice [73]. Such challenges include factors that should be carefully considered during the research phase, issues during the implementation and the translation of the pharmacogenomics findings into clinical practice, and the integration of these findings into public health policy planning.

2.4.1 Ethical and legal issues

A significant body of literature has already addressed the issue of where pharmacogenomics lies within the current ethico-legal frameworks. For example, there are two crucial issues: the increasing and free access to genetic information as well as the secure storage of genetic data. These issues are often likely to hamper advances in pharmacogenomics and add new layers of ethical complexity. Moreover, the ever-existing issues of patients' informed consent combined with the financial cost of pharmacogenomics testing [74] show that there are still critical considerations when integrating genomics into health care.

To this end, current literature has focused on the issues regarding the governance of genomic research and primarily on the informed consent issue. Theoretically, when giving informed consent to research as a participant, adequate and precise information should be given to the individuals to allow an informed and voluntary decision about their participation in the research program. Moreover, the participants should be aware that they can always revoke their decision at any time during the research conduction.

Unfortunately, the theoretically ideal informed consent is rather difficult to achieve in practice. To address this issue, an idea would be to establish a "portable legal consent" (PLC), in which participants would be initially taken through an excessive informative session. Then, the participants would be invited to give their broad consent for their own data to be deposited into a general pool, which will be used exclusively for research purposes under certain and stringent criteria [75]. Stringent criteria should also be implemented for the storage and security of the deposited genetic data to ensure genetic data privacy and the anonymity of the participants.

2.4.2 Raising genomics awareness among healthcare professionals and the general public

Enriching health professionals' genomics knowledge is essential to properly integrate pharmacogenomics into clinical practice. However, recent data show the vast majority of physicians usually lack the appropriate training and understanding of genetics and their contribution towards individualized medicine. This was further supported by data from a US national survey of 10,303 physicians. According to these findings, although most (97.6%) of the participants agreed that genetic variation may affect drug response, only 29% of them had received pharmacogenomics education, whereas only 10.3% of them felt adequately informed about the aspect of pharmacogenomic testing [76].

In another interesting study, Lee et al. [77] demonstrated that the awareness of physicians towards pharmacogenomics and pharmacogenomics testing was significantly improved after receiving their own personalized pharmacogenomics reports using WES data. More precisely, the physicians stated that their perception of the occurrence of adverse drug reactions (ADRs) as well as the contribution of the individual genomic variability to ADRs changed significantly in the post-test survey compared to the pre-testing phase [77].

Previously, Patrinos et al. [73] concluded that the lack of genetics knowledge from health professionals as well as the lack of genomics awareness from both patients and the general public leads to misconception of the benefits of (pharmaco) genetics testing, leading to uncertainly as to which genetic test is better suited for them or, most importantly, has enough scientific evidence to support its clinical utility. These conclusions were supported by research findings from a survey conducted in Greek pharmacists that showed that pharmacists do not feel competent and educated enough to explain the results of pharmacogenomics tests to their clients [78]. This further demonstrates that pharmacists should receive adequate training on the genetic tests and their benefits so as to increase the genomics awareness of their clients. Overall, the findings from these studies highlight the urgent need for improvement of the genetics education that healthcare professionals receive.

2.4.3 Economic evaluation of personalized medicine interventions

The importance of performing pharmacoeconomic studies was highlighted with the upper goal of providing an accurate economic evaluation of pharmacogenomics testing. In particular, we highlighted that there is an urgent need to evaluate the overall cost of pharmacogenomics studies by using different economic evaluation measurements [79]. This way, it will be proven that pharmacogenomics testing is ready for clinical implementation based not only the clinical utility of the current findings but also on the ever-decreasing costs of PGx testing [79].

For example, estimating the cost-effectiveness of pharmacogenomics tests via pharmacoeconomic studies is a necessary and crucial step towards implementation of pharmacogenomics in the clinical routine. Recent pharmacoeconomic studies in cardiovascular disorders [80,81] and cancer [82] have provided sufficient evidence for the cost-effectiveness of pharmacogenomics-guided drug treatments. Based on the findings from these subpopulation-specific pharmacoeconomic studies, favorable economic evaluation results combined with the clinical utility of pre-emptive genotyping are essential factors for the implementation of (pharmaco) genetics testing in clinical practice and reimbursement of the genetic testing costs [74].

Although pharmacogenomics testing will most likely prove to be a cost-effective rather than cost-saving solution, it will greatly improve the management of clinical information thus improving the quality of life of patients and significantly reducing healthcare costs. These aspects will be described in detail in the subsequent chapters of this book.

2.5 Large-scale personalized medicine implementation efforts worldwide

Large-scale nationwide efforts to implement personalized medicine in the clinic are of utmost importance since they demonstrate the feasibility of this concept in terms of clinical implementation while addressing at the same time its public health implications. In this section, we briefly summarize some of the most prominent personalized medicine implementation efforts worldwide.

In the United States, the "All of Us" program (https://allofus.nih.gov), previously known as the "Precision Medicine Initiative," launched by former US President Barack H. Obama in 2015, aims to gather genomic data from one million or more people living in the United States to accelerate research and improve health. As President Obama stated, "...what if figuring out the right dose of medicine was as simple as taking our temperature?"

On a European level, the Ubiquitous Pharmacogenomics Consortium (U-PGx; http://upgx.eu), funded by the European Commission's Horizon-2020 program, is an international effort between universities, academic hospitals, and research institutions to implement pharmacogenomics knowledge in clinical practice [83]. By designing a prospective, block-randomized, controlled clinical study (PREPARE) with 8100 patients, U-PGx aims to reduce the occurrence of severe adverse drug reactions through a cost-effective pharmacogenomics-guided pharmacotherapy. Another example of a large-scale genomics medicine initiative is Genomics England (www.genomicsengland.co.uk), which has begun a very promising genomics medicine project called the 100,000 Genomes Project. In this project, approximately 100,000 UK participants are being sequenced with the goal of improving our genomics and clinical understanding of a variety of diseases (including cancer and rare diseases) and improving patient outcomes.

In 2010, the US National Institutes of Health and the UK Wellcome Trust jointly launched the Human Heredity and Health in Africa (H3Africa) consortium (https://h3africa.org; [84]). H3Africa was formed to catalyze research into genetic diseases in Africa and encourage and facilitate capacity building, namely, infrastructure, resources, training, and ethics. This effort is led by African scientists for the African people and already consists of 48 projects including 34 African countries, leading so far to tangible research (197 scientific articles) and educational (193 PhDs and 127 MSc awarded) deliverables focused on common disorders such as heart and renal disease as well as tuberculosis.

Two other international efforts are the Psychiatric Genetics Consortium (PGC; www.med.unc.edu/pgc/) and the Exome Aggregation Consortium (ExAC; https://gnomad.broadinstitute.org). The PGC focuses on the conduct of meta- and mega-analyses of genome-wide genomic data for psychiatric disorders, thus constituting the largest collaboration effort in the field of psychiatric disorders worldwide. Moreover, the ExAC aggregates and harmonizes exome sequencing data from different large-scale sequencing projects. Its summary data is freely available and in an exportable output form for the scientific community.

In the corporate sector, another example is 23andMe, which is a private personal genomics company (www.23andme.com) well known for their questionable direct-to-consumer genetic testing services. The company has already established numerous academic collaborations and has participated in research projects aiming to identify risk factors for diseases such as Parkinson's disease, lupus, depression, and others. For example, Wray et al. [85], including the 23andMe research team, identified 44 independent and significant loci as risk factors for major depressive disorder. Nevertheless, there are still critical considerations regarding accuracy concerns from consumers as well as regulatory bodies.

Another large-scale genomic medicine initiative comes from AstraZeneca (www.astrazeneca.com), which is focused on revolutionizing drug discovery and development by establishing collaborations with two research institutes (the UK Wellcome Trust Sanger Institute and the Institute for Molecular Medicine in Finland) and a genomic database (Human Longevity, Inc., HLI). Amongst the goals of this initiative is the establishment of an AstraZeneca Centre for Genomic Research as well as development of a database of genome sequence, clinical, and drug-response data, which are donated by half a million patients in clinical trials.

There are also several other pharmacogenomics and genomic medicine implementation initiatives in various other countries worldwide, but their detailed description lies outside the scope of this chapter.

2.6 Conclusions and future perspectives

In this chapter, we summarized some distinct examples of associations of variants within key clinically actionable genes affecting drug efficacy or drug toxicity. These examples have been implemented in the clinical routine and there have been updates in the drug labeling for these associations based on instructions from regulatory agencies (i.e., FDA or EMA). Although massive progress has been made in the field of pharmacogenomics, especially with the technological advance of next-generation sequencing, there are still some critical considerations. More advanced clinical studies combined with the appropriate economic evaluation of pharmacogenomics are central factors that could lead to improvement of clinical pharmacogenomics workflows. Moreover, "genethics" should be carefully considered to raise the awareness of patients and healthcare professional in the concept of personalized medicine.

As such, preemptive pharmacogenomics testing may be a better and more cost- effective approach than the traditional pharmacogenomics approach. By conducting preemptive pharmacogenomics testing, genotyping is simultaneously and prospectively performed for multiple actionable pharmacogenomics biomarkers for each patient, thus making the decision process easier for clinicians and physicians.

In conclusion, many barriers need to be overcome for the proper implementation of pharmacogenomics and personalized medicine interventions in clinical care. However, our growing understanding and knowledge of the genetic bases of response to medications, drug-drug interactions, and variability due to clinical and environmental factors will lead to more widespread use of pharmacogenomics in the near future. Moreover, the growing attention of the pharmaceutical industry and national healthcare policy makers is another beneficial factor that could assist in the precise implementation of personalized medicine in the clinical routine.

Acknowledgments

Part of our own work has been funded by a European Commission grant (H2020-668353; U-PGx) to GPP. The authors declare no conflict of interests.

References

[1] Squassina A, Manchia M, Manolopoulos VG, Artac M, Lappa-Manakou C, Karkabouna S, Mitropoulos K, Del Zompo M, Patrinos GP. Realities and expectations of pharmacogenomics and personalized medicine: impact of translating genetic knowledge into clinical practice. Pharmacogenomics 2010;11:1149–67.

[2] Shen H, Li J, Zhang J, Xu C, Jiang Y, Wu Z, Zhao F, Liao L, Chen J, Lin Y, Tian Q, Papasian CJ, Deng H-W. Comprehensive characterization of human genome variation by high coverage whole-genome sequencing of forty four Caucasians. PLoS One 2013;8(4): e59494.

[3] Piquette-Miller M, Grant DM. The art and science of personalized medicine. Clin Pharmacol Ther 2007;81(3):311–5.

[4] Burstein HJ, Prestrud AA, Seidenfeld J, Anderson H, Buchholz TA, Davidson NE. American society of clinical oncology clinical practice guideline: update on adjuvant endocrine therapy for women with hormone receptor-positive breast cancer. J Clin Oncol 2010;28(23):3784–96.

[5] Borges S, Desta Z, Li L, Skaar TC, Ward BA, Nguyen A, et al. Quantitative effect of CYP2D6 genotype and inhibitors on tamoxifen metabolism: implication for optimization of breast cancer treatment. Clin Pharmacol Ther 2006;80(1):61–74.

[6] Goetz MP, Rae JM, Suman VJ, Safgren SL, Ames MM, Visscher DW, et al. Pharmacogenetics of tamoxifen biotransformation is associated with clinical outcomes of efficacy and hot flashes. J Clin Oncol 2005;23 (36):9312–8.

[7] Schroth W, Antoniadou L, Fritz P, Schwab M, Muerdter T, Zanger UM, et al. Breast cancer treatment outcome with adjuvant tamoxifen relative to patient **CYP2D6** and CYP2C19 genotypes. J Clin Oncol 2007;25 (33):5187–93.

[8] Goetz MP, Kamal A, Ames MM. Tamoxifen pharmacogenomics: the role of **CYP2D6** as a predictor of drug response. Clin Pharmacol Ther 2008;83(1):160–6.

[9] Iyer L, Hall D, Das S, Mortell MA, Ramírez J, Kim S, et al. Phenotype-genotype correlation of in vitro SN-38 (active metabolite of irinotecan) and bilirubin glucuronidation in human liver tissue with UGT1A1 promoter polymorphism. Clin Pharmacol Ther 1999;65(5):576–82.

[10] Ando Y, Saka H, Ando M, Sawa T, Muro K, Ueoka H, Yokoyama A, Saitoh S, Shimokata K, Hasegawa Y. Polymorphisms of UDP-glucuronosyltransferase gene and irinotecan toxicity: a pharmacogenetic analysis. Cancer Res 2000;60(24):6921–6.

[11] Iyer L, Das S, Janisch L, Wen M, Ramírez J, Karrison T, et al. UGT1A1*28 polymorphism as a determinant of irinotecan disposition and toxicity. Pharmacogenomics J 2002;2(1):43–7.

[12] Innocenti F, Undevia SD, Iyer L, Chen PX, Das S, Kocherginsky M, Karrison T, Janisch L, Ramírez J, Rudin CM, Vokes EE, Ratain MJ. Genetic variants in the UDP-glucuronosyltransferase 1A1 gene predict the risk of severe neutropenia of irinotecan. J Clin Oncol 2004;22(8):1382–8.

[13] Kim TW, Innocenti F. Insights, challenges, and future directions in irinogenetics. Ther Drug Monit 2007;29 (3):265–70.

[14] Pandey K, Dubey RS, Prasad BB. A critical review on clinical application of separation techniques for selective recognition of uracil and 5-fluorouracil. Indian J Clin Biochem 2016;31(1):3–12.

[15] Lunenburg CA, Henricks LM, Guchelaar HJ, Swen JJ, Deenen MJ, Schellens JH, Gelderblom H. Prospective DPYD genotyping to reduce the risk of fluoropyrimidine-induced severe toxicity: ready for prime time. Eur J Cancer 2016;54(2):40–8.

[16] Deenen MJ, Meulendijks D, Cats A, Sechterberger MK, Severens JL, Boot H, et al. Upfront genotyping of DPYD*2A to individualize fluoropyrimidine therapy: a safety and cost analysis. J Clin Oncol 2016;34 (3):227–34.

[17] Terrazzino S, Cargnin S, Del Re M, Danesi R, Canonico PL, Genazzani AA. DPYD IVS14+1G>A and 2846A>T genotyping for the prediction of severe fluoropyrimidine-related toxicity: a meta-analysis. Pharmacogenomics 2013;14(11):1255–72.

[18] Meulendijks D, Henricks LM, Sonke GS, Deenen MJ, Froehlich TK, Amstutz U, et al. Clinical relevance of DPYD variants c.1679T>G, c.1236G>A/HapB3, and c.1601G>A as predictors of severe fluoropyrimidine-associated toxicity: a systematic review and meta-analysis of individual patient data. Lancet Oncol 2015;16 (16):1639–50.

[19] Rosmarin D, Palles C, Pagnamenta A, Kaur K, Pita G, Martin M, et al. A candidate gene study of capecitabine-related toxicity in colorectal cancer identifies new toxicity variants at DPYD and a putative role for ENOSF1 rather than TYMS. Gut 2015;64(1):111–20.

[20] Henricks LM, Lunenburg CA, Meulendijks D, Gelderblom H, Cats A, Swen JJ, et al. Translating DPYD genotype into DPD phenotype: using the DPYD gene activity score. Pharmacogenomics 2015;16(11):1277–86.

[21] Van Kuilenburg AB, Meinsma R, Zoetekouw L, Van Gennip AH. High prevalence of the IVS14 + 1G>A mutation in the dihydropyrimidine dehydrogenase gene of patients with severe 5-fluorouracil-associated toxicity. Pharmacogenetics 2002;12(7):555–8.

[22] Kitzmiller JP, Groen DK, Phelps MA, Sadee W. Pharmacogenomic testing: relevance in medical practice: why drugs work in some patients but not in others. Cleve Clin J Med 2011;78(4):243–57.

[23] Hulot JS, Bura A, Villard E, Azizi M, Remones V, Goyenvalle C, Aiach M, Lechat P, Gaussem P. Cytochrome P450 2C19 loss-of-function polymorphism is a major determinant of clopidogrel responsiveness in healthy subjects. Blood 2006;108:2244–7.

[24] Yin T, Miyata T. Pharmacogenomics of clopidogrel: evidence and perspectives. Thromb Res 2011; 128:307–16.

[25] Bouman HJ, Schömig E, van Werkum JW, Velder J, Hackeng CM, Hirschhäuser C, Waldmann C, Schmalz H, ten Berg JM, Taubert D. Paraoxonase-1 is a major determinant of clopidogrel efficacy. Nat Med 2011;17 (1):110–6.
[26] Simon T, Verstuyft C, Mary-Krause M, Quteineh L, Drouet E, Meneveau N, Steg PG, Ferrieres J, Danchin N, Becquemont L. Genetic determinants of response to clopidogrel and cardiovascular events. N Engl J Med 2009;360:363–75.
[27] Brown SA, Pereira N. Pharmacogenomic impact of CYP2C19 variation on clopidogrel therapy in precision cardiovascular medicine. J Pers Med 2018;8(1):E8.
[28] Rieder MJ, Reiner AP, Gage BF, Nickerson DA, Eby CS, McLeod HL, Blough DK, Thummel KE, Veenstra DL, Rettie AE. Effect of VKORC1 haplotypes on transcriptional regulation and warfarin dose. N Engl J Med 2005;352(22):2285–93.
[29] Takeuchi F, McGinnis R, Bourgeois S, Barnes C, Eriksson N, Soranzo N, Whittaker P, Ranganath V, Kumanduri V, McLaren W, Holm L, Lindh J, Rane A, Wadelius M, Deloukas P. A genome-wide association study confirms VKORC1, CYP2C9, and CYP4F2 as principal genetic determinants of warfarin dose. PLoS Genet 2009;5, e1000433.
[30] Manolopoulos VG, Ragia G, Tavridou A. Pharmacogenetics of coumarinic oral anticoagulants. Pharmacogenomics 2010;11(4):493–6.
[31] Perez-Andreu V, Roldan V, Lopez-Fernandez MF, Anton AI, Alberca I, Corral J, Montes R, Garcia-Barbera N, Ferrando F, Vicente V, Gonzalez-Conejero R. Pharmacogenetics of acenocoumarol in patients with extreme dose requirements. J Thromb Haemost 2010;8:1012–7.
[32] Donnelly LA, Doney AS, Tavendale R, Lang CC, Pearson ER, Colhoun HM, McCarthy MI, Hattersley AT, Morris AD, Palmer CN. Common nonsynonymous substitutions in slco1b1 predispose to statin intolerance in routinely treated individuals with type 2 diabetes: a go-darts study. Clin Pharmacol Ther 2011;89:210–6.
[33] Voora D, Shah SH, Spasojevic I, Ali S, Reed CR, Salisbury BA, Ginsburg GS. The SLCO1B1*5 genetic variant is associated with statin-induced side effects. J Am Coll Cardiol 2009;54:1609–16.
[34] Kitzmiller JP, Binkley PF, Pandey SR, Suhy AM, Baldassarre D, Hartmann K. Statin pharmacogenomics: pursuing biomarkers for predicting clinical outcomes. Discov Med 2013;16(86):45–51.
[35] Li Y, Iakoubova OA, Shiffman D, Devlin JJ, Forrester JS, Superko HR. KIF6 polymorphism as a predictor of risk of coronary events and of clinical event reduction by statin therapy. Am J Cardiol 2010;106(7):994–8.
[36] Papp AC, Pinsonneault JK, Wang D, Newman LC, Gong Y, Johnson JA, Pepine CJ, Kumari M, Hingorani AD, Talmud PJ. Cholesteryl Ester Transfer Protein (CETP) polymorphisms affect mRNA splicing, HDL levels, and sex-dependent cardiovascular risk. PLoS One 2012;7(3), e31930.
[37] Aral H, Vecchio-Sadus A. Toxicity of lithium to humans and the environment—a literature review. Ecotoxicol Environ Saf 2008;70(3):349–56.
[38] Yatham LN, Kennedy SH, Parikh SV, Schaffer A, Beaulieu S, Alda M, O'Donovan C, Macqueen G, McIntyre RS, Sharma V, Ravindran A, Young LT, Milev R, Bond DJ, Frey BN, Goldstein BI, Lafer B, Birmaher B, Ha K, Nolen WA, Berk M. Canadian Network for Mood and Anxiety Treatments (CANMAT) and International Society for Bipolar Disorders (ISBD) collaborative update of CANMAT guidelines for the management of patients with bipolar disorder: update 2013. Bipolar Disord 2013;15(1):1–44.
[39] Pisanu C, Heilbronner U, Squassina A. The role of pharmacogenomics in bipolar disorder: moving towards precision medicine. Mol Diagn Ther 2018;22(4):409–20.
[40] Amare AT, Schubert KO, Klingler-Hoffmann M, Cohen-Woods S, Baune BT. The genetic overlap between mood disorders and cardiometabolic diseases: a systematic review of genome wide and candidate gene studies. Transl Psychiatry 2017;7(1), e1007.
[41] Song J, Bergen SE, Di Florio A, Karlsson R, Charney A, Ruderfer DM, Stahl EA, Members of the International Cohort Collection for Bipolar Disorder (ICCBD), Chambert KD, Moran JL, Gordon-Smith K, Forty L, Green EK, Jones I, Jones L, Scolnick EM, Sklar P, Smoller JW, Lichtenstein P, Hultman C, Craddock N, Landén M, Smoller JW, Perlis RH, Lee PH, Castro VM, Hoffnagle AG, Sklar P, Stahl EA, Purcell SM,

Ruderfer DM, Charney AW, Roussos P, Michele Pato CP, Medeiros H, Sobel J, Craddock N, Jones I, Forty L, Florio AD, Green E, Jones L, Gordon-Smith K, Landen M, Hultman C, Jureus A, Bergen S, McCarroll S, Moran J, Smoller JW, Chambert K, Belliveau RA. Genome-wide association study identifies SESTD1 as a novel risk gene for lithium-responsive bipolar disorder. Mol Psychiatry 2016;21(9):1290–7.

[42] Squassina A, Manchia M, Borg J, Congiu D, Costa M, Georgitsi M, Chillotti C, Ardau R, Mitropoulos K, Severino G, Del Zompo M, Patrinos GP. Evidence for association of an ACCN1 gene variant with response to lithium treatment in Sardinian patients with bipolar disorder. Pharmacogenomics 2011;12:1559–69.

[43] Xu Q, Wu X, Li M, Huang H, Minica C, Yi Z, Wang G, Shen L, Xing Q, Shi Y, He L, Qin S. Association studies of genomic variants with treatment response to risperidone, clozapine, quetiapine and chlorpromazine in the Chinese Han population. Pharm J 2016;16(4):357–65.

[44] Hamilton SP. The promise of psychiatric pharmacogenomics. Biol Psychiatry 2015;77(1):29–35.

[45] Pouget JG, Gonçalves VF, Nurmi EL, Laughlin C, Mallya KS, McCracken JT, Aman MG, McDougle CJ, Scahill L, Misener VL, Tiwari AK, Zai CC, Brandl EJ, Felsky D, Leung AQ, Lieberman JA, Meltzer HY, Potkin SG, Niedling C, Steimer W, Leucht S, Knight J, Müller DJ, Kennedy JL. Investigation of TSPO variants in schizophrenia and antipsychotic treatment outcomes. Pharmacogenomics 2015;16(1):5–22.

[46] Blasi G, Selvaggi P, Fazio L, Antonucci LA, Taurisano P, Masellis R, Romano R, Mancini M, Zhang F, Caforio G, Popolizio T, Apud J, Weinberger DR, Bertolino A. Variation in dopamine D2 and serotonin 5-HT2A receptor genes is associated with working memory processing and response to treatment with antipsychotics. Neuropsychopharmacology 2015;40:1600–8.

[47] Hwang R, Tiwari AK, Zai CC, Felsky D, Remington E, Wallace T, Tong RP, Souza RP, Oh G, Potkin SG, Lieberman JA, Meltzer HY, Kennedy JL. Dopamine D4 and D5 receptor gene variant effects on clozapine response in schizophrenia: replication and exploration. Prog Neuro-Psychopharmacol Biol Psychiatry 2012;37:62–75.

[48] Yu H, Yan H, Wang L, Li J, Tan L, Deng W, Chen Q, Yang G, Zhang F, Lu T, Yang J, Li K, Lv L, Tan Q, Zhang H, Xiao X, Li M, Ma X, Yang F, Li L, Wang C, Li T, Zhang D, Yue W. Five novel loci associated with antipsychotic treatment response in patients with schizophrenia: a genome-wide association study. Lancet Psychiatry 2018;5:327–38.

[49] Amberger JS, Bocchini CA, Schiettecatte F, Scott AF, Hamosh A. OMIM.org: Online Mendelian Inheritance in Man (OMIM®), an online catalog of human genes and genetic disorders. Nucleic Acids Res 2015;43: D789–98.

[50] Giardine B, van Baal S, Kaimakis P, Riemer C, Miller W, Samara M, Kollia P, Anagnou NP, Chui DH, Wajcman H, Hardison RC, Patrinos GP. HbVar database of human hemoglobin variants and thalassemia mutations: 2007 update. Hum Mutat 2007;28:206.

[51] Hardison RC, Chui DH, Giardine B, Riemer C, Patrinos GP, Anagnou N, Miller W, Wajcman H. HbVar: a relational database of human hemoglobin variants and thalassemia mutations at the globin gene server. Hum Mutat 2002;19:225–33.

[52] Patrinos GP, Wajcman H. Recording human globin gene variation. Hemoglobin 2004;28(2):v–vii.

[53] Lek M, Karczewski KJ, Minikel EV, Samocha KE, Banks E, Fennell T, O'Donnell-Luria AH, Ware JS, Hill AJ, Cummings BB, Tukiainen T, Birnbaum DP, Kosmicki JA, Duncan LE, Estrada K, Zhao F, Zou J, Pierce-Hoffman E, Berghout J, Cooper DN, Deflaux N, DePristo M, Do R, Flannick J, Fromer M, Gauthier L, Goldstein J, Gupta N, Howrigan D, Kiezun A, Kurki MI, Moonshine AL, Natarajan P, Orozco L, Peloso GM, Poplin R, Rivas MA, Ruano-Rubio V, Rose SA, Ruderfer DM, Shakir K, Stenson PD, Stevens C, Thomas BP, Tiao G, Tusie-Luna MT, Weisburd B, Won HH, Yu D, Altshuler DM, Ardissino D, Boehnke M, Danesh J, Donnelly S, Elosua R, Florez JC, Gabriel SB, Getz G, Glatt SJ, Hultman CM, Kathiresan S, Laakso M, McCarroll S, McCarthy MI, McGovern D, McPherson R, Neale BM, Palotie A, Purcell SM, Saleheen D, Scharf JM, Sklar P, Sullivan PF, Tuomilehto J, Tsuang MT, Watkins HC, Wilson JG, Daly MJ, MacArthur DG, Exome Aggregation Consortium. Analysis of protein-coding genetic variation in 60,706 humans. Nature 2016;536:285–91.

[54] Karczewski KJ, Francioli LC, Tiao G, Cummings BB, Alföldi J, Wang Q, Collins RL, Laricchia KM, Ganna A, Birnbaum DP, Gauthier LD, Brand H, Solomonson M, Watts NA, Rhodes D, Singer-Berk M, Seaby EG, Kosmicki JK, Walters RK, Tashman K, Farjoun Y, Banks E, Poterba T, Wang A, Seed C, Whiffin N, Chong JX, Samocha KE, Pierce-Hoffman E, Zappala Z, O'Donnell-Luria AH, Minikel EV, Weisburd B, Lek M, Ware JS, Vittal C, Armean IM, Bergelson L, Cibulskis K, Connolly KM, Covarrubias M, Donnelly S, Ferriera S, Gabriel S, Gentry J, Gupta N, Jeandet T, Kaplan D, Llanwarne K, Munshi R, Novod S, Petrillo N, Roazen D, Ruano-Rubio V, Saltzman A, Schleicher M, Soto J, Tibbetts K, Tolonen C, Wade G, Talkowski ME, The Genome Aggregation Database Consortium, Neale BM, Daly MJ, MacArthur DJ. Variation across 141,456 human exomes and genomes reveals the spectrum of loss-of-function intolerance across human protein-coding genes. bioRxiv 2019. https://doi.org/10.1101/531210.

[55] Sherry ST, Ward MH, Kholodov M, Baker J, Phan L, Smigielski EM, Sirotkin K. dbSNP: the NCBI database of genetic variation. Nucleic Acids Res 2001;29(1):308–11.

[56] Viennas E, Komianou A, Mizzi C, Stojiljkovic M, Mitropoulou C, Muilu J, Vihinen M, Grypioti P, Papadaki S, Pavlidis C, Zukic B, Katsila T, van der Spek PJ, Pavlovic S, Tzimas G, Patrinos GP. Expanded national database collection and data coverage in the FINDbase worldwide database for clinically relevant genomic variation allele frequencies. Nucleic Acids Res 2017;45(D1):D846–53.

[57] Landrum MJ, Lee JM, Riley GR, Jang W, Rubinstein WS, Church DM, Maglott DR. ClinVar: public archive of relationships among sequence variation and human phenotype. Nucleic Acids Res 2014;42(Database issue):D980–5.

[58] Hamosh A, Scott AF, Amberger JS, Bocchini CA, McKusick VA. Online Mendelian Inheritance in Man (OMIM), a knowledgebase of human genes and genetic disorders. Nucleic Acids Res 2005;33:D514–7.

[59] Stenson PD, Ball EV, Mort M, Phillips AD, Shiel JA, Thomas NS, Abeysinghe S, Krawczak M, Cooper DN. Human gene mutation database (HGMD): 2003 update. Hum Mutat 2003;21(6):577–81.

[60] den Dunnen JT, Dalgleish R, Maglott DR, Hart RK, Greenblatt MS, McGowan-Jordan J, Roux AF, Smith T, Antonarakis SE, Taschner PE. HGVS recommendations for the description of sequence variants: 2016 update. Hum Mutat 2016;37(6):564–9.

[61] Fokkema IF, Taschner PE, Schaafsma GC, Celli J, Laros JF, Den Dunnen JT. LOVD v.2.0: the next generation in gene variant databases. Hum Mutat 2011;32(5):557–63.

[62] Wright ES. Using DECIPHER v2.0 to analyze big biological sequence data in R. R J 2016;8.

[63] O'Leary NA, Wright MW, Brister JR, Ciufo S, Haddad D, McVeigh R, Rajput B, Robbertse B, Smith-White B, Ako-Adjei D, Astashyn A, Badretdin A, Bao Y, Blinkova O, Brover V, Chetvernin V, Choi J, Cox E, Ermolaeva O, Farrell CM, Goldfarb T, Gupta T, Haft D, Hatcher E, Hlavina W, Joardar VS, Kodali VK, Li W, Maglott D, Masterson P, McGarvey KM, Murphy MR, O'Neill K, Pujar S, Rangwala SH, Rausch D, Riddick LD, Schoch C, Shkeda A, Storz SS, Sun H, Thibaud-Nissen F, Tolstoy I, Tully RE, Vatsan AR, Wallin C, Webb D, Wu W, Landrum MJ, Kimchi A, Tatusova T, DiCuccio M, Kitts P, Murphy TD, Pruitt KD. Reference sequence (RefSeq) database at NCBI: current status, taxonomic expansion, and functional annotation. Nucleic Acids Res 2016;44(D1):D733–45.

[64] MacArthur JA, Morales J, Tully RE, Astashyn A, Gil L, Bruford EA, Larsson P, Kileck P, Dalgleish R, Maglott DR, Cunningham F. Locus Reference Genomic: reference sequences for the reporting of clinically relevant sequence variants. Nucleic Acids Res 2014;42:D873–8.

[65] Lott MT, Leipzig JN, Derbeneva O, Xie HM, Chalkia D, Sarmady M, Procaccio V, Wallace DC. mtDNA variation and analysis using MITOMAP and MITOMASTER. Curr Protoc Bioinformatics 2013;1(123). 1.23.1-26.

[66] Whirl-Carrillo M, McDonagh EM, Hebert JM. Pharmacogenomics knowledge for personalized medicine. Clin Pharmacol Ther 2012;92:414–7.

[67] Relling M, Klein T. CPIC: clinical pharmacogenetics implementation consortium of the pharmacogenomics research network. Clin Pharmacol Ther 2011;89:464–7.

[68] Flicek P, Aken BL, Beal K, Ballester B, Caccamo M, Chen Y, Clarke L, Coates G, Cunningham F, Cutts T, Down T, Dyer SC, Eyre T, Fitzgerald S, Fernandez-Banet J, Gräf S, Haider S, Hammond M, Holland R,

Howe KL, Howe K, Johnson N, Jenkinson A, Kähäri A, Keefe D, Kokocinski F, Kulesha E, Lawson D, Longden I, Megy K, Meidl P, Overduin B, Parker A, Pritchard B, Prlic A, Rice S, Rios D, Schuster M, Sealy I, Slater G, Smedley D, Spudich G, Trevanion S, Vilella AJ, Vogel J, White S, Wood M, Birney E, Cox T, Curwen V, Durbin R, Fernandez-Suarez XM, Herrero J, Hubbard TJ, Kasprzyk A, Proctor G, Smith J, Ureta-Vidal A, Searle S. Ensembl 2008. Nucleic Acids Res 2008;36:D707–14.

[69] Karolchik D, Kuhn RM, Baertsch R, Barber GP, Clawson H, Diekhans M, Giardine B, Harte RA, Hinrichs AS, Hsu F, Kober KM, Miller W, Pedersen JS, Pohl A, Raney BJ, Rhead B, Rosenbloom KR, Smith KE, Stanke M, Thakkapallayil A, Trumbower H, Wang T, Zweig AS, Haussler D, Kent WJ. The UCSC genome browser database: 2008 update. Nucleic Acids Res 2008;36:D773–9.

[70] Wheeler DL, Barrett T, Benson DA, Bryant SH, Canese K, Chetvernin V, Church DM, DiCuccio M, Edgar R, Federhen S, Geer LY, Kapustin Y, Khovayko O, Landsman D, Lipman DJ, Madden TL, Maglott DR, Ostell J, Miller V, Pruitt KD, Schuler GD, Sequeira E, Sherry ST, Sirotkin K, Souvorov A, Starchenko G, Tatusov RL, Tatusova TA, Wagner L, Yaschenko E. Database resources of the National Center for Biotechnology Information. Nucleic Acids Res 2007;35:D5–D12.

[71] Lathe W, Williams J, Mangan M, Karolchik D. Genomic data resources: challenges and promises. Nat Educ 2008;1(3):2.

[72] Giannopoulou E, Katsila T, Mitropoulou C, Tsermpini E-E, Patrinos GP. Integrating next-generation sequencing in the clinical pharmacogenomics workflow. Front Pharmacol 2019;10:384.

[73] Patrinos GP, Baker DJ, Al-Mulla F, Vasiliou V, Cooper DN. Genetic tests obtainable through pharmacies: the good, the bad, and the ugly. Hum Genom 2013;7:17.

[74] Symeonidis S, Koutsilieri S, Vozikis A, Cooper DN, Mitropoulou C, Patrinos GP. Application of economic evaluation to assess feasibility for reimbursement of genomic testing as part of personalized medicine interventions. Front Pharmacol 2019;10:830.

[75] Vayena E, Prainsack B. Regulating genomics: time for a broader vision. Sci Transl Med 2013;5:198ed12.

[76] Stanek EJ, Sanders CL, Taber KA, Khalid M, Patel A, Verbrugge RR, Agatep BC, Aubert RE, Epstein RS, Frueh FW. Adoption of pharmacogenomic testing by US physicians: results of a nationwide survey. Clin Pharmacol Ther 2012;91:450–8.

[77] Lee KH, Min BJ, Kim JH. Personal genome testing on physicians improves attitudes on pharmacogenomic approaches. PLoS One 2019;14(3), e0213860.

[78] Mai Y, Mitropoulou C, Papadopoulou XE, Vozikis A, Cooper DN, van Schaik RH, Patrinos GP. Critical appraisal of the views of healthcare professionals with respect to pharmacogenomics and personalized medicine in Greece. Perinat Med 2014;11:15–26.

[79] Patrinos GP, Mitropoulou C. Measuring the value of pharmacogenomics evidence. Clin Pharmacol Ther 2017;102(5):739–41.

[80] Fragoulakis V, Bartsakoulia M, Díaz-Villamarín X, Chalikiopoulou K, Kehagia K, Ramos JGS, Martínez-González LJ, Gkotsi M, Katrali E, Skoufas E, Vozikis A, John A, Ali BR, Wordsworth S, Dávila-Fajardo CL, Katsila T, Patrinos GP, Mitropoulou C. Cost-effectiveness analysis of pharmacogenomics-guided clopidogrel treatment in Spanish patients undergoing percutaneous coronary intervention. Pharm J 2019;19.

[81] Mitropoulou C, Fragoulakis V, Bozina N, Vozikis A, Supe S, Bozina T, Poljakovic Z, van Schaik RH, Patrinos GP. Economic evaluation for pharmacogenomic-guided warfarin treatment for elderly Croatian patients with atrial fibrillation. Pharmacogenomics 2015;16(2):137–48.

[82] Fragoulakis V, Roncato R, Fratte CD, Ecca F, Bartsakoulia M, Innocenti F, Toffoli G, Cecchin E, Patrinos GP, Mitropoulou C. Estimating the effectiveness of DPYD genotyping in Italian individuals suffering from cancer based on the cost of chemotherapy-induced toxicity. Am J Hum Genet 2019;104(6):1158–68.

[83] van der Wouden CH, Cambon-Thomsen A, Cecchin E, Cheung K-C, Dávila-Fajardo CL, Deneer VH, Dolžan V, Ingelman-Sundberg M, Jönsson S, Karlsson MO, Kriek M, Mitropoulou C, Patrinos GP, Pirmohamed M, Samwald M, Schaeffeler E, Schwab M, Steinberger D, Stingl J, Sunder-Plassmann G, Toffoli G, Turner RM, van Rhenen MH, Swen JJ, Guchelaar H-J, On behalf of the Ubiquitous Pharmacogenomics Consortium.

Implementing pharmacogenomics in Europe: design and implementation strategy of the ubiquitous pharmacogenomics consortium. Clin Pharmacol Ther 2017;101(3):341–58.

[84] H3Africa Consortium. Research capacity. Enabling the genomic revolution in Africa. Science 2014;344 (6190):1346–8.

[85] Wray NR, Ripke S, Mattheisen M, Trzaskowski M, Byrne EM, Abdellaoui A, Adams MJ, Agerbo E, Air TM, Andlauer TMF, Bacanu SA, Bækvad-Hansen M, Beekman AFT, Bigdeli TB, Binder EB, Blackwood DRH, Bryois J, Buttenschøn HN, Bybjerg-Grauholm J, Cai N, Castelao E, Christensen JH, Clarke TK, Coleman JIR, Colodro-Conde L, Couvy-Duchesne B, Craddock N, Crawford GE, Crowley CA, Dashti HS, Davies G, Deary IJ, Degenhardt F, Derks EM, Direk N, Dolan CV, Dunn EC, Eley TC, Eriksson N, Escott-Price V, Kiadeh FHF, Finucane HK, Forstner AJ, Frank J, Gaspar HA, Gill M, Giusti-Rodríguez P, Goes FS, Gordon SD, Grove J, Hall LS, Hannon E, Hansen CS, Hansen TF, Herms S, Hickie IB, Hoffmann P, Homuth G, Horn C, Hottenga JJ, Hougaard DM, Hu M, Hyde CL, Ising M, Jansen R, Jin F, Jorgenson E, Knowles JA, Kohane IS, Kraft J, Kretzschmar WW, Krogh J, Kutalik Z, Lane JM, Li Y, Li Y, Lind PA, Liu X, Lu L, MacIntyre DJ, MacKinnon DF, Maier RM, Maier W, Marchini J, Mbarek H, McGrath P, McGuffin P, Medland SE, Mehta D, Middeldorp CM, Mihailov E, Milaneschi Y, Milani L, Mill J, Mondimore FM, Montgomery GW, Mostafavi S, Mullins N, Nauck M, Ng B, Nivard MG, Nyholt DR, O'Reilly PF, Oskarsson H, Owen MJ, Painter JN, Pedersen CB, Pedersen MG, Peterson RE, Pettersson E, Peyrot WJ, Pistis G, Posthuma D, Purcell SM, Quiroz JA, Qvist P, Rice JP, Riley BP, Rivera M, Saeed Mirza S, Saxena R, Schoevers R, Schulte EC, Shen L, Shi J, Shyn SI, Sigurdsson E, Sinnamon GBC, Smit JH, Smith DJ, Stefansson H, Steinberg S, Stockmeier CA, Streit F, Strohmaier J, Tansey KE, Teismann H, Teumer A, Thompson W, Thomson PA, Thorgeirsson TE, Tian C, Traylor M, Treutlein J, Trubetskoy V, Uitterlinden AG, Umbricht D, Van der Auwera S, van Hemert AM, Viktorin A, Visscher PM, Wang Y, Webb BT, Weinsheimer SM, Wellmann J, Willemsen G, Witt SH, Wu Y, Xi HS, Yang J, Zhang F, eQTLGen, 23andMe, Arolt V, Baune BT, Berger K, Boomsma DI, Cichon S, Dannlowski U, de Geus JEC, DePaulo JR, Domenici E, Domschke K, Esko T, Grabe HJ, Hamilton SP, Hayward C, Heath AC, Hinds DA, Kendler KS, Kloiber S, Lewis G, Li QS, Lucae S, Madden PFA, Magnusson PK, Martin NG, McIntosh AM, Metspalu A, Mors O, Mortensen PB, Müller-Myhsok B, Nordentoft M, Nöthen MM, O'Donovan MC, Paciga SA, Pedersen NL, Penninx BWJH, Perlis RH, Porteous DJ, Potash JB, Preisig M, Rietschel M, Schaefer C, Schulze TG, Smoller JW, Stefansson K, Tiemeier H, Uher R, Völzke H, Weissman MM, Werge T, Winslow AR, Lewis CM, Levinson DF, Breen G, Børglum AD, Sullivan PF, Major Depressive Disorder Working Group of the Psychiatric Genomics Consortium. Genome-wide association analyses identify 44 risk variants and refine the genetic architecture of major depression. Nat Genet 2018;50:668–81.

CHAPTER

Economic evaluation of genome-guided treatment in oncology

3

Paula K. Lorgelly

School of Population Health and Department of Economics, University of Auckland, Auckland, New Zealand

3.1 Introduction

In 2020 there were an estimated 19.3 million new cases of cancer and 10.0 million deaths globally [1]. Cancer is the leading cause of death worldwide after cardiovascular disease [2]. Currently, the burden of cancer is borne by high-income countries. For example, 23% of cancer cases are in Europe, including 20% of cancer deaths [3], yet Europe only has 9% of the global population; however, low- and middle-income countries are facing an increasing burden [4]. Many cancer deaths can be prevented by modifying or avoiding risk factors like smoking and alcohol consumption. Screening and vaccination programs are also effective interventions to reduce the burden.

For those with a cancer diagnosis, there is a range of treatment modalities available, including surgery, radiotherapy, and pharmacological treatments. Until two decades ago, most of the pharmacological therapies used in cancer treatment were chemotherapeutic antitumor drugs, which work by killing replicating cancer cells, but chemotherapy can also kill normal non-cancerous cells. The late 1990s was a period of significant evolution in cancer treatments with the development of targeted therapies (also referred to in the literature and clinical practice as personalized medicines or precision medicines). These treatments target specific genes, proteins, or the tissue environment to control the growth, division, and spread of cancer cells.

Targeted therapies work using different mechanisms. They can inhibit growth signals, as is the case with the first targeted oncology drug to be approved, rituximab. Other drugs exert their effect by inhibiting angiogenesis, where tumors are blocked from making the new blood vessels they need to keep growing; this is the mechanism of action for bevacizumab, first approved for colorectal cancer and subsequently approved for use in lung, breast, renal, brain, and ovarian cancer (it is also used in age-related macular degeneration). In addition, some drugs act by inducing apoptosis, where cells with DNA too damaged to repair, such as cancer cells, can be forced to die; bortezomib does this in lymphoma and multiple myeloma [5]. These drugs can take the form of monoclonal antibodies (MABs) that block a specific target on the outside of cancer cells and/or target the area around the cancer, which includes delivering a drug or radioactive substance directly to cancer cells. Alternatively, they may take the form of a small-molecule drug that blocks the process that helps cancer cells multiply and spread, such as angiogenesis inhibitors. Note that cancer vaccines and gene therapies are sometimes also considered targeted therapies because they interfere with the growth of specific cancer cells.

Economic Evaluation in Genomic and Precision Medicine. https://doi.org/10.1016/B978-0-12-813382-8.00009-4

Whether targeted therapy is indicated, and if so, which targeted therapy is appropriate, is determined by genomic tests. These tests can identify gene mutations that may be influencing a cancer's behavior. Heterogeneity in a tumor means that individuals with the same cancer may not have the same mutations; conversely, different cancers can have the same mutation. For example, about half of all melanomas have changes (mutations) in the BRAF gene. These can often be successfully treated with vemurafenib, dabrafenib, and encorafenib, all of which target the BRAF protein. These drugs are not likely to work in individuals whose melanomas have a normal BRAF gene. A small proportion (4%) of non-small cell lung cancers (NSCLCs) have also been found to have changes in the BRAF gene; dabrafenib has been used successfully to target this mutation.

The US Food and Drug Administration (FDA) has approved more than 100 different targeted therapies in more than fifteen types of cancers [6,7]. Research and development is such that we are now in an era of tumor-agnostic or site-agnostic treatments, where a drug is used to treat any kind of cancer, regardless of where in the body it started or the type of tissue from which it developed. Pembrolizumab, a PD-1 inhibitor, was the first drug to be approved with a tumor-agnostic indication. It can be used to treat adult and pediatric patients with unresectable or metastatic solid tumors irrespective of the site of the cancer, as long as the tumor has been identified as having microsatellite instability-high (MSI-H) or deficient DNA mismatch repair (dMMR).

Testing technology has also advanced over time. Individual biomarker assays have given way to next-generation sequencing (NGS)-based gene panel tests. NGS-based panels not only detect multiple gene alterations but can also simultaneously detect other structural variants. Large genomic studies like The Cancer Genome Atlas [8] and the International Cancer Genome Consortium [9] utilized NGS, and in doing so identified a number of new and unknown targets, leading to several new therapies. Although the cost of NGS has declined over time, and these tests are now being used more frequently in clinical practice, questions remain regarding whether NGS tests are as affordable as $1000 per genome [10,11]. Diagnostic tests can also be co-developed with targeted treatment and are subsequently known as companion diagnostics (defined by the FDA as a "tool that provides information that is essential for the safe and effective use of a corresponding therapeutic product"). Companion diagnostics commonly aim to identify a single biomarker to align with a single drug in a linear development approach; however, panel testing platforms are emerging in the form of next-generation companion diagnostics [12].

As a result of unprecedented research and development, healthcare professionals now have a variety of diagnostics with which to identify targets, and numerous drugs at their disposal with which to target the cancer therapy. But not all targeted treatments are effective (and although targeted they can still cause serious side effects) or cost-effective. Similarly, not all tests offer the same clinical utility or offer value for money, and these issues can be further amplified when it is a companion diagnostic that is being assessed. For patients to benefit from targeted treatments and for this benefit not to be at the expense of other patients (i.e., to minimize the opportunity cost), it is important that genome-guided treatments deliver value for money. Several issues arise when assessing these targeted therapies. This chapter provides an overview of the methodological challenges that accompany the assessment of the cost-effectiveness of genome-guided treatments in oncology. It specifically considers the intertwined relationship between test and therapy and how that is modeled, the value of testing beyond traditional health outcomes and quality adjusted life years (QALYs), and the challenges tumor-agnostic therapies and their testing regimes bring. It also considers the role of economic evaluation in reimbursement decisions, both in terms of pricing the test and the treatment, particularly where a treatment has multiple indications.

3.2 Methodological challenges

Evaluating targeted therapies challenges the standard application of economic evaluation methodologies. Many of the challenges are with respect to the evaluation of the testing regime, including the interrelationship between the test and the treatment, which we discuss in detail later in the chapter. Yang et al. [13] reviewed Health Technology Assessment (UK NIHR HTA) reports that evaluated diagnostic tests using decision analytic modeling. Overall, most models passed their quality checklist, but there were areas where methodological improvements were warranted. Many studies failed to systematically identify treatment effects, comparator selection was often poorly described, and several HTA reports assumed independence of tests used in sequence. In a similar vein, Doble et al. reviewed economic evaluations of companion diagnostics used in oncology [14]. They found considerable variation in the application of methods for incorporating test-related into model-based evaluations of targeted therapies. A key issue was the limited inclusion of test sensitivity and specificity, and the impact of ignoring incorrect diagnoses on cost-effectiveness (this is discussed further later in the chapter). Additional methodological issues highlighted in the review included how the timing and sequence of multiple tests were modeled, how to adjust for variable or varying test clinical thresholds, and whether and how the cost of the test is included in the analysis [14].

These are just two papers that highlight the breadth of methodological challenges and how they result in considerable variability in the application of economic evaluation methods to diagnostics. Notably, both papers developed quality assessment methodological checklists for diagnostics, thereby providing decision makers and commissioners who utilize economic evidence with tools to differentiate high- and low-quality economic evaluations of diagnostic tests. Checklists aside, some methodological challenges are enduring, including the intertwined relationship between test and therapy.

3.3 The inseparability of test and treatment

The inseparability of the value of the test and the value of the treatment, and specifically the value that the test brings to the treatment, is often ignored when assessing the cost-effectiveness of targeted treatment. This is despite the fact that the success or otherwise of targeted treatments depends on the presence or absence of the target and therefore on the test's performance, that is, its sensitivity or specificity [15]. Dichotomizing the test result into positive and negative and restricting any assessment of a treatment's effectiveness and cost-effectiveness to positive results is misleading. This is in part due to the sensitivity and specificity of the test; individuals who receive a positive diagnostic test result will incur the cost of the test and the intervention but will only accrue the health benefits of the intervention if the test result is a true-positive. Individuals with a false-positive result will incur the cost of the test and the intervention but may experience a net health decline due to side effects because in the absence of the target no health improvement would be expected [15]. Similarly, those who receive a negative test will incur the cost of the test, but the consequences of a true-negative result will differ from those with a false-negative result.

Excluding the costs of testing by not explicitly examining the means of identifying targets is unfortunately common. This means that when a targeted test is expensive and when tests have variable sensitivity and specificity, ignoring negative test results can introduce considerable bias. Such an analysis will also exclude the negative health consequences of failing to give the intervention to those with a false-negative result.

Elkin et al. [15], in a review of previous economic evaluations that evaluated the cost-effectiveness of trastuzumab used for advanced HER2-positive cancer, found that many studies did not explicitly consider the HER2 testing strategies. Most of the studies ignored the cost of testing, while others assigned a total cost to testing HER2-positive patients but did not consider test performance or the outcomes of patients who tested negative. The consequences of such omissions are neatly described by Kurian et al.: [16] "Our analysis is based on the assumption that the HER2/neu status of the patient is already known ... If we had included HER2/neu testing costs, ... the ICER for trastuzumab-based therapies would be less favorable" (p. 639).

A possible reason for ignoring the testing component of targeted treatment could be that HTA agencies have well-developed systems for evaluating pharmaceuticals but not diagnostics [17] or, if they do evaluate diagnostics, there are often separate methods (and committees) for evaluating them, which may introduce inconsistencies [18]. The International Society for Pharmacoeconomics and Outcomes Research (ISPOR) Devices and Diagnostics Special Interest Group reviewed diagnostic-specific HTA programs [17] and found that for the few HTA agencies that have diagnostic-specific methodologies/guidelines there was:

(1) No clear mandate as to which diagnostics need formal HTA.
(2) No uniform approach for laboratory-developed tests (e.g., whether they should be formally evaluated by HTA agencies along with regulatory-approved tests, or whether payers should consider them differently with regard to pricing and reimbursement).
(3) No clear guidance regarding how outcomes should be measured, appropriate study types, performance requirements, comparative effectiveness, and economic threshold.
(4) Variable and unclear impact of HTA recommendations on reimbursement, access, and pricing.
(5) Limited criteria assessed to translate the HTA findings into molecular diagnostic pricing and reimbursement decision making.

The National Institute for Health and Care Excellence (NICE) provides a good example of this heterogeneity within a single agency [19]. Companion diagnostics that are co-developed with a therapy are often reviewed within the technology appraisal programme (TAP), but if the marketing authorization does not require the use of the companion diagnostic then these diagnostics can be appraised via the diagnostics assessment programme (DAP), as would be the case for other targeted tests that are not companions. The TAP process considers how companion diagnostics support optimal use of the pharmaceutical but does not necessarily include a detailed evaluation of the companion diagnostic or alternative companion diagnostic options. Such a detailed evaluation is the remit of the DAP, but if a diagnostic is evaluated under the TAP it won't also be evaluated under the DAP. Notably, only recommendations made through TAP and the highly specialized technologies programme (HSTP) have mandatory funding status. Thus, if a diagnostic is evaluated via the DAP and considered effective and cost-effective it is not necessarily funded by National Health Service (NHS) Trusts. While targeted tests and targeted therapies are intrinsically inseparable, decision making often treats them as distinct separate entities.

3.4 Value of testing

The inseparability of test and treatment has fueled a debate on the value of diagnostics. As Soares et al. [20] argue, the mechanism of value for diagnostic and prognostic technologies is more complex than that for other health technologies. While they may affect health directly (perhaps by lessening anxiety), the main value of a test typically lies in its ability to identify patients who will benefit from treatment; "the mechanism by which value is generated is not direct, but instead arises from tailoring treatment decisions to patient characteristics" (p. 496). Notably, Soares et al. sought to define a framework for establishing the value of diagnostic and prognostic tests in a way that is consistent with methods used for the evaluation of other health technologies. Others have instead argued that the value or benefits of testing need to be evaluated outside of our current methodological paradigm, as such tests offer benefits that are beyond traditional health outcomes [21–23].

Grosse et al. [21] provide a useful overview of the options that are available to value the outcomes of genetic tests that allow the evaluation to extend beyond cost-effectiveness analysis. They discuss the limitations of cost-effectiveness analyses and the focus on maximizing health, particularly in terms of measuring QALY gains with respect to diagnostic testing applications. They argue that cost benefit analysis would provide an estimate of the willingness to pay that reflects the monetary valuation of diagnostic outcomes and encompasses both health and non-health attributes.

There have been several attempts to value diagnostic tests in the broader sense. Most of these studies have undertaken discrete choice experiments to explore trade-offs between different attributes, using a cost attribute to estimate the willingness to pay for diagnostic tests. For example, Weymann et al. [22] elicited preferences from patients to understand the value of using massively parallel sequencing (MPS) to assess colorectal cancer (CRC) risk. They found that patients value information on the genetic causes of CRC and suggest that replacing traditional diagnostic testing with MPS testing would increase patients' utility. Buchanan et al. [23] also found that patients (in this instance those with chronic lymphocytic leukemia) value the information genomic tests provide on which individuals may not respond to therapy as well as value process attributes like the time it takes to turn around a test result. A discrete choice experiment was also used to elicit preferences for a complex genomic sequencing (CGS) technology in members of the public and patients with advanced cancer as part of the iPREDICT Melbourne Genomics Health Alliance study [24]. Similar process attributes were found to be valued by respondents, such as the timeliness of receiving test results, the location in which the results are given to patients, and the expertise of the healthcare professional. These are all healthcare features that would not necessarily deliver a health outcome. This is further confirmed in another component of the study that assessed the health-related quality of life (HRQoL) of patients using the EQ-5D-5L at baseline (patient undergoes CGS test) and during follow-up (including reporting of results). In this analysis, the EQ-5D-5L was unable to pick up any changes in HRQoL, whereas preferences elicited at baseline and follow-up showed that patients place a value on the technology.

Such preference elicitation studies confirm the value of genomic-guided therapy and in particular the value of targeted testing, but several challenges remain. One issue is whether preference elicitation exercises should be undertaken with patients or the general public [25]. Experience plays an important role in the formulation of preferences; patients are better informed and more able to make trade-offs while considering factors such as adaptation [26]. However, the opportunity cost argument would suggest that the preferences of the general public—those who pay into the healthcare system and bear the cost of funding one technology and not another—are paramount. The debate on this continues [27].

Even if this debate were resolved, a key remaining challenge is the inseparability of testing and treatment and the need to consider their value jointly. Spackman et al. [28] argue that the current HTA paradigm (use of cost-effectiveness analysis) would be appropriate if a broad measure of outcome that reflects all relevant health and non-health consequences existed, or there was some acceptable way to explicitly trade off distinct outcomes (e.g., how much population health should be forgone to generate information of non-health value to individuals). Currently, neither of these exist.

3.5 Cost-effectiveness comparator

A final methodological challenge that will likely gain importance given the growing pan-tumor pipeline [29] relates to the comparator employed in an economic evaluation of genome-guided tumor-agnostic therapies. Tumor-agnostic therapies are argued to be a game-changer with respect to how we treat cancer, but their benefits will only be realized by patients with cancer if regulatory and reimbursement agencies approve them. Regulatory agencies have demonstrated an openness to a labeling approach that recognizes the histologically independent nature of these therapies [30]. However, it is important to remember that regulatory agencies assess safety and efficacy, not comparative effectiveness, and economic evidence is also required to support approval.

Traditionally, a treatment's safety and efficacy are assessed in as large a segment of the patient population that can feasibly be recruited to a study. Targeted therapy challenges this paradigm, as the patient population with the appropriate biological profile for such a study is often small. Trial designs are adapted and so-called master protocols are employed that allow for multiple targeted agents to be tested in relatively small patient cohorts. Tumor-agnostic therapies challenge this even further as clinical trials must span multiple histologies. Basket trials, typically used as phase II screening trials for the off-label use of a targeted drug in patients with the same genomic alterations for which it was approved, are viewed as an efficient method of generating evidence of the efficacy of tumor-agnostic therapies [31]. However, HTA agencies generally favor randomized control trials with straightforward designs. They also require that evidence reviews include an active comparator that is most relevant to their healthcare setting, which will further challenge data collection and analysis. An additional issue is that the evidence base may lack information on the relationship between biomarker-based endpoints and overall survival, which devalues the level of evidence presented to an HTA panel [30]. This uncertainty is exacerbated by small sample sizes for rare cancers (if they share biomarkers with common cancers), different prognoses for different cancers, and differences in unmet need. Overall, this makes the assessment of comparative effectiveness very difficult.

It is important to acknowledge that some tumor-agnostic treatments have been approved by HTA agencies. NICE recently recommended larotrectinib for use within the Cancer Drugs Fund (CDF), although its inclusion in the CDF indicates that its cost-effectiveness is uncertain, requiring the collection of additional evidence. Khogeer et al. [32] describe additional issues related to tumor-agnostic therapies and uncertainty, noting that the size of the eligible patient population often remains unknown, given constant advances in identifying biomarkers in new tumor histologies. Commissioners may therefore be concerned about the future budget impact of treatments targeting tumor-agnostic indications.

3.6 Pricing and reimbursement

It would be remiss to discuss economic evaluation without touching on the role it plays in pricing and reimbursement [33], particularly given concerns regarding the budget impact of genome-guided treatment in cancer. A technology is considered to be cost-effective and offering value for money if the value it delivers in terms of health benefits is greater than the health it displaces [34]. This happens when the incremental cost effectiveness ratio (ICER) is less than the decision threshold. When price is aligned with patient benefit this is effectively value-based pricing (VBP). Since the initial discussions of using VBP in the UK NHS much has been written and debated [35–38], and more recently this debate has considered the role of VBP for cures and in orphan diseases [39,40]. Some of this debate centers on what is valued, the decision threshold for these values, and the role of affordability. These are also issues for targeted therapies, but two further challenges exist: VBP of indications and VBP of diagnostics.

In a recent paper discussing pricing of precision medicines Tiriveedhi [41] argued that pricing of laboratory diagnostics is cost-based, whilst drug pricing is value-based. Consider these in turn, starting with the latter. Many countries with functioning HTA systems have cost-effectiveness thresholds and thus employ a VBP approach to decision making, where value is commonly defined using QALYs. In these systems, however, reimbursement or pricing models are often based on a "one price for one drug" rule. This means that if a drug has multiple indications, then irrespective of the indication and, importantly, irrespective of the effectiveness of the drug in that indication, they face the same market price (give or take discounts, rebates, or weighted average prices). A single price for a single drug creates a disconnect between price and value, and in doing so creates distorted incentives [42]. If products are priced according to the indication (or use where value differs) and not the drug itself then that would be referred to as indication-based pricing (IBP) (also known as indication-specific pricing (ISP) or multi-indication pricing (MIP)). IBP allows the price to be linked with value, and as such is a type of VBP. It is an efficient approach to segmenting the market and ensuring all patients who can benefit from innovative medicines do benefit, but at a fair, transparent value-driven price [42].

Targeted therapies often have multiple indications as they target genetic mutations; only recently have tumor-agnostic treatments received approval, that is, been approved irrespective of their indication. While targeted therapies may have multiple indications (based on the site or stage of the cancer) this does not mean they are equally as effective in each indication (this is also the case for tumor-agnostic treatments; effectiveness and comparative effectiveness are likely to vary when evaluated by site/stage). Trastuzumab is a good example of this; it first gained marketing approval in the United Kingdom for use in advanced HER2-positive breast cancer in 2000 and received a NICE recommendation in 2002. More recently, trastuzumab has received approval for use in advanced gastric cancer. It is effective, but its comparative effectiveness is lower compared to its use in breast cancer. The application of IBP should therefore result in a lower price when used in gastric cancer compared with breast cancer [43]. However, what often happens is that the price for all indications changes to reflect a weighted average price or blended price, or indication-specific risk-sharing (or managed entry) agreements are used [44]. The lack of IBP and VBP for targeted treatments may mean that drug developers are not incentivized to launch specific indications (perhaps those with low value but still beneficial) or they may sequence indication launches according to commercial interests rather than to address unmet patient needs. Evidence of this, however, is scarce, in part due to the commercial-in-confidence nature of drug development. See Box 3.1 for an example of how IBP can influence launch sequence.

BOX 3.1 Effect of single (lowest) price on launch sequence

Consider a targeted therapy that is first launched in indication A, then indication B (which is of higher incremental value and has a larger patient population), then in indication C (which has the highest value but the smallest patient population), and finally in indication D (which has a patient population of A, B, and C combined but the lowest incremental benefit).

Under a pricing model that applies a single price as the lowest price (Fig. A), the price would be P1 from the launch of indication A, and remain at P1 for the launch of indications B and C. This may mean a manufacturer is reluctant to launch indications B and C, as the price does not reflect their value. This is most probable for indication C, as it has the smallest patient population despite it having the greatest benefit. If the manufacturer launches indication D, then the price will fall to P2, which will then erode the returns on all previous indications launched.

If possible, a manufacturer may change their launch sequence (Fig. B). Launching first with the highest value indication (indication C) and achieving P1, then launching indication B with a price reduction to P2. The manufacturer may not launch indications A and D because they face further price reductions. As a result, many patients with high unmet need are denied (or delayed) access to an efficacious drug.

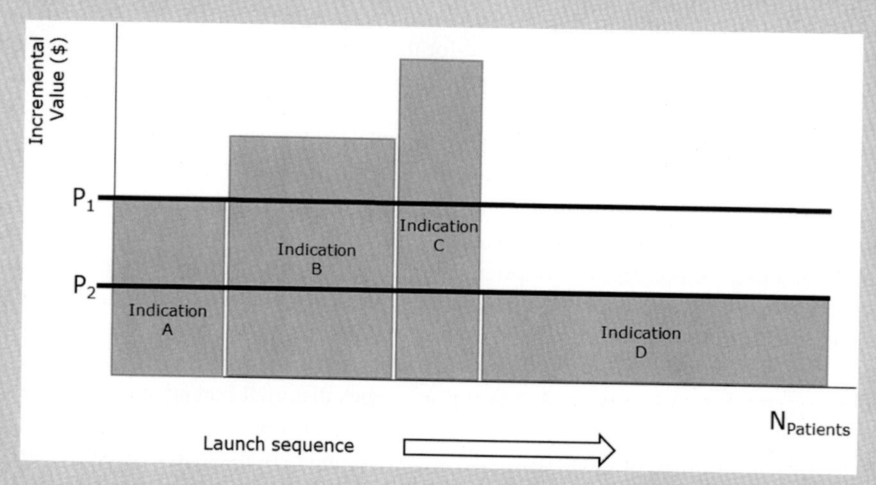

FIG. A

Single lowest price.

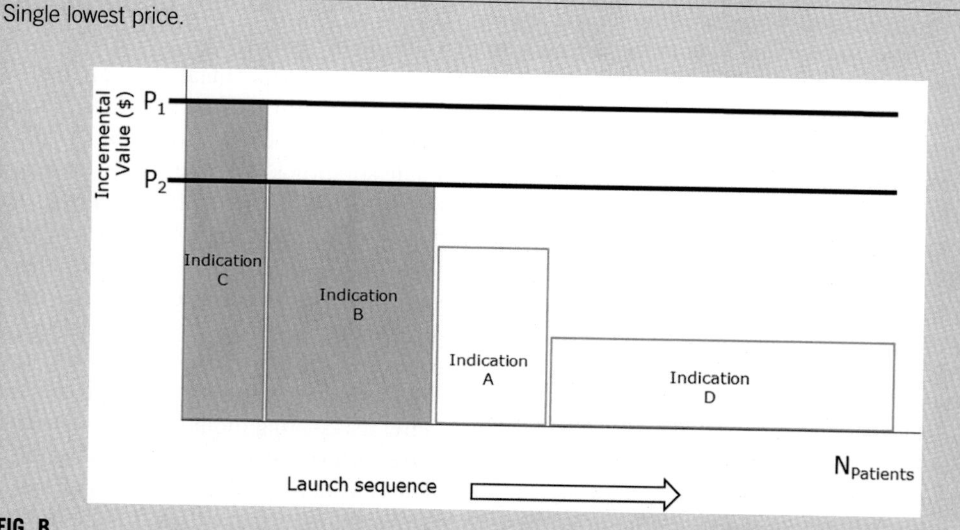

FIG. B

Single lowest price with sequencing.

Testing for targets further complicates pricing and reimbursement. As Tiriveedhi [41] noted, diagnostics are often priced at cost not value; indeed in the early stages of market access pharmaceutical companies generally provide the diagnostic test free of charge. But diagnostics have value in themselves, particularly in the age of NGS-based gene panels. This raises two questions: what is the value of a diagnostic that on its own does not deliver health gains per se, just information, and how do you differentiate the value of a diagnostic from the value of the treatment and therefore apply VBP?

As noted, economic evaluations of diagnostic tests regularly employ a QALY framework as dictated by HTA agencies. Annemans et al. [45] argue that QALYs do not sufficiently cover all patient benefits and that companion diagnostics and other targeted tests offer considerable process utility. Process utility is the satisfaction resulting from the process of providing a technology rather than the outcome of the technology [46]. Other researchers have gone further and argued that complementary diagnostics (which include companion diagnostics) offer value in knowing by reducing uncertainty, giving hope, having real option value and insurance value, and providing scientific spillovers [47]. Metrics to quantify, measure, and value these are limited [48].

Even if we could measure these broader value elements, we are still challenged by the optimal division of rewards to ensure innovation is rewarded and promoted in both the diagnostic and therapeutic areas. Philipson [49] suggests that for companion diagnostics the issue is analogous to a two-part pricing model: the price of the diagnostic is the entry fee and the price of the therapy is the cost of a ride (as per Oi's Disneyland pricing model [50]). This assumes joint ownership. For example, Roche Diagnostics has a portfolio of assays to detect the HER2 biomarker thereby identifying patients eligible for the targeted Roche drug Herceptin (trastuzumab). However, while many large pharmaceutical companies have diagnostic divisions, partnerships between diagnostics manufacturers and pharmaceutical companies are more common. This is particularly true for NGS-based panels that are a growth market [51]. Key to this partnership is the division of value.

Garrison and Towse [52] argue that rewards should be split to encourage dynamic efficiency (see Box 3.2 for their example). To understand how value is apportioned differently between tests and therapies consider two examples provided by Garrison and Towse [52]:

(1) Chronic myeloid leukemia (CML) is caused by one translocation that creates a singular mutation: the BCR-ABL fusion gene or Philadelphia chromosome. This discovery focused research on therapies to block the BCR-ABL gene. Imatinib blocks this gene, so arguably much of the value or the benefit created is derived from the drug rather than the diagnosis of the gene mutation.

(2) Oncotype Dx is a genetic test that determines how a tumor will behave and respond to treatment. Its value and (economic) benefit are realized due the avoidance of chemotherapy costs and side effects. For this reason, it provides considerable value and can therefore command a relatively high VBP given it avoids unproductive chemotherapy.

The value of targeted therapy is only recognized once a diagnostic can identify appropriate patients. However, given the research and development and approval processes are different in terms of timing, evidence generation, the nature of the evidence, and market access approvals, there is a need for greater synergies in order to realize the value of genome-guided treatment in oncology. A commonality is that reimbursement incentivizes innovation for both pharmaceutical and diagnostic manufacturers. Economic evaluations and improving the methodological quality of economic evaluations should contribute to achieving fair value-driven prices and a fair process to attribute value, thereby ensuring continued innovation in targeted therapies.

BOX 3.2 Dividing the combination value between the drug and the diagnostic

Consider a hypothetical case where an existing biopharmaceutical is on the market and earning $1 billion in annual revenues despite having only a 50% response rate. Suppose a reliable biomarker-based test is invented that can predict the responders.

In theory, the payer should be indifferent to giving the Dx manufacturer $500 million and perhaps a premium for reducing the uncertainty for both the payer and the patient.

Although it is important to remember that it was the drug manufacturer's invention that created the health gain in the responders.

The extra value created by the Dx manufacturer consists of (1) avoiding any adverse event treatment costs and any related health losses in the non-responders plus (2) the "value-of-knowing premium" (if the premium were, say, 5% of total health gain, it would be worth $50 million).

Assuming flexible VBP, the drug innovator's price would be roughly doubled (i.e., to $1 billion in revenues) and the Dx innovator would receive a reward in relation to cost savings and QALY gains in the non-responders plus a premium for the value of knowing (i.e., $50 million).

Adapted from Garrison L, Towse A. Personalized medicine: pricing and reimbursement policies as a potential barrier to development and adoption, economics of. Encyclopedia of health economics, 2014.

3.7 Conclusion

There are several challenges to evaluating the efficiency of genome-guided therapies in oncology. Some of these are unique to cancer care, but there are several commonalities that will resonate in other diseases discussed elsewhere in this publication. The interdependence of genomic testing and the use (or not) of therapies is a crucial element. It is important to model test accuracy, as this influences the success or otherwise of the therapy. As noted, this is often overlooked and as such economic evaluations are therefore not of a sufficient quality to inform decisions about the adoption of targeted therapies. Even when correctly modeled it is important to reflect on the fact that there is other information beyond test sensitivity and specificity that patients value, such as the process and environment a test is delivered in. To date, however, the way we measure outcomes means these considerations are often overlooked. However, as NGS-based gene panels become increasingly common, offering information not just on a target of interest but multiple genetic mutations and other variants, we will need to develop new ways to measure health and non-health outcomes. Likewise, we will need to consider the applicability of HTA methods for assessing the cost-effectiveness of tumor-agnostic targeted therapies.

These limitations in our economic evaluation toolkit should not be considered as criticisms. Scientific innovation will lead to methods innovation, which, if this leads to innovations in pricing and reimbursement approaches, should complete a virtuous cycle: manufacturers are appropriately incentivized to develop innovative targeted technologies and collate evidence of efficacy and comparative effectiveness, thereby ensuring that HTA appraisal will be evidence-based and adopted technologies fairly reimbursed. Until then, to quote Philips et al. [53]: "Health economists have two key challenges … to continue to apply robust methods of economic evaluation … [and] to tackle the methodological and practical issues to generate a sufficient evidence base to inform resource allocation decisions" (p. 123).

References

[1] Sung H, Ferlay J, Siegel RL, Laversanne M, Soerjomataram I, Jemal A, Bray F. Global cancer statistics 2020: GLOBOCAN estimates of incidence and mortality worldwide for 36 cancers in 185 countries. CA Cancer J Clin 2021;71(3):209–49.

[2] Roth GA, Abate D, Abate KH, Abay SM, Abbafati C, Abbasi N, et al. Global, regional, and national age-sex-specific mortality for 282 causes of death in 195 countries and territories, 1980–2017: a systematic analysis for the Global Burden of Disease Study 2017. Lancet 2018;392(10159):1736–88.

[3] Ferlay J, Colombet M, Soerjomataram I, Dyba T, Randi G, Bettio M, et al. Cancer incidence and mortality patterns in Europe: estimates for 40 countries and 25 major cancers in 2018. Eur J Cancer 2018;103:356–87.

[4] World Health Organization. Global Health Observatory (GHO) data. NCD mortality and morbidity, 2022. Available from: https://www.who.int/gho/ncd/mortality_morbidity/en/.

[5] Hassan M, Watari H, AbuAlmaaty A, Ohba Y, Sakuragi N. Apoptosis and molecular targeting therapy in cancer. Biomed Res Int 2014;2014.

[6] Kumar GL. FDA-approved targeted therapies in oncology. In: Badve S, Kumar GL, editors. Predictive biomarkers in oncology: applications in precision medicine. Cham: Springer International Publishing; 2019. p. 605–22.

[7] NIH National Cancer Institute. Targeted cancer therapies, 2020. Available from: https://www.cancer.gov/about-cancer/treatment/types/targeted-therapies/targeted-therapies-fact-sheet.

[8] Tomczak K, Czerwińska P, Wiznerowicz M. The Cancer Genome Atlas (TCGA): an immeasurable source of knowledge. Contemp Oncol 2015;19(1A):A68.

[9] Zhang J, Baran J, Cros A, Guberman JM, Haider S, Hsu J, et al. International Cancer Genome Consortium Data Portal—a one-stop shop for cancer genomics data. Database 2011;2011.

[10] Phillips KA, Pletcher MJ, Ladabaum U. Is the "$1000 Genome" really $1000? Understanding the full benefits and costs of genomic sequencing. Technol Health Care 2015;23(3):373.

[11] Schwarze K, Buchanan J, Fermont JM, Dreau H, Tilley MW, Taylor JM, et al. The complete costs of genome sequencing: a microcosting study in cancer and rare diseases from a single center in the United Kingdom. Genet Med 2020;22(1):85–94.

[12] Khoury JD, Catenacci DV. Next-generation companion diagnostics: promises, challenges, and solutions. Arch Pathol Lab Med 2015;139(1):11–3.

[13] Yang Y, Abel L, Buchanan J, Fanshawe T, Shinkins B. Use of decision modelling in economic evaluations of diagnostic tests: an appraisal and review of health technology assessments in the UK. Pharmacoecon Open 2019;1–11.

[14] Doble B, Tan M, Harris A, Lorgelly P. Modeling companion diagnostics in economic evaluations of targeted oncology therapies: systematic review and methodological checklist. Expert Rev Mol Diagn 2015;15(2): 235–54.

[15] Elkin EB, Marshall DA, Kulin NA, Ferrusi IL, Hassett MJ, Ladabaum U, et al. Economic evaluation of targeted cancer interventions: critical review and recommendations. Genet Med 2011;13(10):853–60.

[16] Kurian AW, Thompson RN, Gaw AF, Arai S, Ortiz R, Garber AM. A cost-effectiveness analysis of adjuvant trastuzumab regimens in early HER2/neu-positive breast cancer. J Clin Oncol 2007;25(6):634–41.

[17] Garfield S, Polisena J, Spinner DS, Postulka A, Lu CY, Tiwana SK, et al. Health technology assessment for molecular diagnostics: practices, challenges, and recommendations from the medical devices and diagnostics special interest group. Value Health 2016;19(5):577–87.

[18] Gallego G, Harris A. Evaluation in a disconnected healthcare system: problems and suggested solutions from the Australian HTA review. Expert Rev Pharm Outcomes Res 2010;10(6):615–7.

[19] Cowles E, Marsden G, Cole A, Devlin N. A review of NICE methods and processes across health technology assessment programmes: why the differences and what is the impact? Appl Health Econ Health Policy 2017;15(4):469–77.

[20] Soares MO, Walker S, Palmer SJ, Sculpher MJ. Establishing the value of diagnostic and prognostic tests in Health Technology Assessment. Med Decis Mak 2018;38(4):495–508.
[21] Grosse SD, Wordsworth S, Payne K. Economic methods for valuing the outcomes of genetic testing: beyond cost-effectiveness analysis. Genet Med 2008;10(9):648–54.
[22] Weymann D, Veenstra DL, Jarvik GP, Regier DA. Patient preferences for massively parallel sequencing genetic testing of colorectal cancer risk: a discrete choice experiment. Eur J Hum Genet 2018;26(9):1257–65.
[23] Buchanan J, Wordsworth S, Schuh A. Patients' preferences for genomic diagnostic testing in chronic lymphocytic leukaemia: a discrete choice experiment. Patient 2016;9(6):525–36.
[24] Smith K, O'Haire S, Khuong-Quang D-A, Markman B, Gan H, Ekert PG, et al, editors. iPREDICT: incorporating complex profiling of patients to enrol onto molecularly directed cancer therapeutics. Preliminary results of the adult and paediatric. Asia-Pacific J Clin Oncol 2019;15.
[25] Versteegh M, Brouwer W. Patient and general public preferences for health states: a call to reconsider current guidelines. Soc Sci Med 2016;165:66–74.
[26] Helgesson G, Ernstsson O, Åström M, Burström K. Whom should we ask? A systematic literature review of the arguments regarding the most accurate source of information for valuation of health states. Qual Life Res 2020;1-18.
[27] Nicolet A, van Asselt AD, Vermeulen KM, Krabbe PF. Value judgment of new medical treatments: Societal and patient perspectives to inform priority setting in The Netherlands. PLoS One 2020;15(7), e0235666.
[28] Spackman E, Hinde S, Bojke L, Payne K, Sculpher M. Using cost-effectiveness analysis to quantify the value of genomic-based diagnostic tests: recommendations for practice and research. Genet Test Mol Biomarkers 2017;21(12):705–16.
[29] Looney A, Nawaz K, Webster R. Tumour-agnostic therapies. Nat Rev Drug Discov 2020;19(6):383–4.
[30] Thomas M, Vora D, Schmidt H. Preparing health systems for tumour-agnostic treatment; 2019.
[31] Dittrich C. Basket trials: from tumour gnostic to tumour agnostic drug development. Cancer Treat Rev 2020;, 102082.
[32] Khogeer B, Anjarwalla N, Doolub N, Grosvenor A. PHP206-tumour-agnostic agents: are they fit for reimbursement? Value Health 2018;21:S184–5.
[33] Drummond M, Jönsson B, Rutten F. The role of economic evaluation in the pricing and reimbursement of medicines. Health Policy 1997;40(3):199–215.
[34] Claxton K, Martin S, Soares M, Rice N, Spackman E, Hinde S, et al. Methods for the estimation of the National Institute for Health and Care Excellence cost-effectiveness threshold. Health Technol Assess 2015;19(14):1–503. https://doi.org/10.3310/hta19140.
[35] Claxton K. Oft, Vbp: Qed? Health Econ 2007;16(6):545–58.
[36] Claxton K, Briggs A, Buxton MJ, Culyer AJ, McCabe C, Walker S, et al. Value based pricing for NHS drugs: an opportunity not to be missed? BMJ 2008;336(7638):251–4.
[37] Hughes DA. Value-based pricing. Springer; 2011.
[38] Towse A. If it ain't broke, don't price fix it: the OFT and the PPRS. Health Econ 2007;16(7):653–65.
[39] Danzon PM. Affordability challenges to value-based pricing: mass diseases, orphan diseases, and cures. Value Health 2018;21(3):252–7.
[40] Pearson SD, Ollendorf DA, Chapman RH. New cost-effectiveness methods to determine value-based prices for potential cures: what are the options? Value Health 2019;22(6):656–60.
[41] Tiriveedhi V. Impact of precision medicine on drug repositioning and pricing: a too small to thrive crisis. J Personal Med 2018;8(4):36.
[42] Cole A, Towse A, Lorgelly P, Sullivan R. Economics of innovative payment models compared with single pricing of pharmaceuticals. OHE Res Paper 2018;18:4.
[43] Yeung K, Li M, Carlson JJ. Using performance-based risk-sharing arrangements to address uncertainty in indication-based pricing. J Managed Care Special Pharm 2017;23(10):1010–5.

[44] Pearson SD, Dreitlein WB, Henshall C, Towse A. Indication-specific pricing of pharmaceuticals in the US healthcare system. J Comp Effect Res 2017;6(5):397–404.
[45] Annemans L, Redekop K, Payne K. Current methodological issues in the economic assessment of personalized medicine. Value Health 2013;16(6):S20–6.
[46] Donaldson C, Shackley P. Does "process utility" exist? A case study of willingness to pay for laparoscopic cholecystectomy. Soc Sci Med 1997;44(5):699–707.
[47] Garrison L, Mestre-Ferrandiz J, Zamora B. The value of knowing and knowing the value: improving the health technology assessment of complementary diagnostics. London: Office of Health Economics, EPEMED; 2016.
[48] Lakdawalla DN, Doshi JA, Garrison Jr LP, Phelps CE, Basu A, Danzon PM. Defining elements of value in health care—a health economics approach: an ISPOR Special Task Force report [3]. Value Health 2018;21(2):131–9.
[49] Philipson TJ. The economic value and pricing of personalized medicine. Econ Dimen Personal Precision Med 2018;9–19.
[50] Oi WY. A Disneyland dilemma: two-part tariffs for a Mickey Mouse monopoly. Q J Econ 1971;85(1):77–96.
[51] Phillips KA, Douglas MP. The global market for next-generation sequencing tests continues its torrid pace. J Precision Med 2018;4.
[52] Garrison L, Towse A. Personalized medicine: pricing and reimbursement policies as a potential barrier to development and adoption, economics of. In: Encyclopedia of health economics; 2014.
[53] Phillips KA, Payne K, Redekop K. Personalized medicine: economic evaluation and evidence. In: World scientific handbook of global health economics and public policy: volume 2: health determinants and outcomes. World Scientific; 2016. p. 123–50.

CHAPTER

Economic evaluation of rare diseases and the diagnostic odyssey

4

Dean A. Regier[a,b], Deirdre Weymann[a], Ian Cromwell[a], Morgan Ehman[a], and Samantha Pollard[a]

[a]*Cancer Control Research, BC Cancer, Vancouver, BC, Canada,* [b]*School of Population and Public Health, University of British Columbia, Vancouver, BC, Canada*

4.1 Introduction

There are 6000–8000 known rare diseases, each affecting fewer than 8 in 10,000 people. When combined, 1 in 12 people is affected by rare diseases with two-thirds being children [1]. Rare diseases are often complex, life-threatening, or chronically debilitating and can only be managed with the combined efforts of multiple healthcare professionals [2]. When drugs are available to treat rare diseases, they are often very expensive, with recent drugs costing up to $300,000 per patient [3].

More than 80% of rare diseases are believed to have a genetic origin [4]. Traditional genetic testing for rare diseases can involve a time-consuming, multistep process of sequentially interrogating single genes until an etiologic diagnosis is achieved. This diagnostic odyssey—the time and healthcare resources used between symptom onset and definitive diagnosis—varies considerably across patients [5]. Up to 30% of patients with a rare disease report waiting 5 years or longer to receive a diagnosis [6]. On average, patients consult five different hospital specialists and receive three misdiagnoses before the cause of their rare disease is determined [7].

Over the past decade, technological innovation has led to genome-wide approaches with greater resolution compared to traditional testing [8]. These tests include chromosomal microarrays (CMA) and high-throughput next-generation sequencing (NGS) approaches. Although genomic tests have shown promise for diagnosis and treatment for rare diseases, their implementation into healthcare systems is varied [9]. This can be attributed in part to the high cost of NGS technologies and the limited clinical and economic evidence base supporting adoption. Within the context of constrained healthcare budgets, decision-makers require evidence of effectiveness and cost-effectiveness. In this chapter, we discuss the use of genomics to diagnose rare diseases and the challenges of this application for economic evaluations. We then explore how health economists have addressed challenges to date, provide recommendations moving forward, and highlight areas where additional research is needed.

4.2 Genomics and rare diseases

Rare diseases are a heterogeneous group of disorders classified based on their low prevalence within the general population. This classification is contextual, and diseases considered rare vary across jurisdictions. Prevalence thresholds for identifying rare diseases typically range from about 1 to 8 in

Economic Evaluation in Genomic and Precision Medicine. https://doi.org/10.1016/B978-0-12-813382-8.00001-X

10,000 [10]. Prevalence estimates can be poorly defined due to limited epidemiological data and diseases may transition to and from this classification.

Though individually rare, the aggregate burden of rare diseases is significant. Approximately 6% to 8% of the global population is affected by a rare disease, and the majority of these conditions have a genetic etiology [11]. Contributing to this burden is the chronic, multisystem, progressive, and degenerative nature of these diseases. Rare diseases often begin as pediatric disorders [12,13]. Up to 35% of deaths in the first year of life are due to these rare conditions and quality of life impacts in children who survive to adulthood can be severe [14,15]. Curative treatments do not exist for the majority of rare diseases [16]. Even when treatment is available, it is often prohibitively expensive [17]. This financial impact is substantial when considered alongside the indirect costs of rare diseases, with some affected individuals reporting average annual out-of-pocket costs of $17,000 per patient [18].

4.2.1 Diagnostic odyssey

Traditional diagnostic testing for rare diseases can involve a series of single-gene tests that are often inconclusive. Up to 50% of patients with a rare disorder go undiagnosed [19]. Without a diagnosis, patients and families experience a diagnostic odyssey involving inconclusive specialist visits, diagnostic investigations, and potential misdiagnoses.

The goal of traditional diagnostic testing is to reach a point when a confirmatory genetic test can be conducted (Fig. 4.1). To achieve this goal, a parsimonious candidate gene list and effective single-gene tests are required to identify pathogenic variants. When candidate genes for a disease are few and well defined, this method can be highly effective. In the context of idiopathic disorders or when there is considerable genetic heterogeneity, the sequential testing approach can be inefficient.

The heterogeneity of rare diseases poses challenges for traditional diagnostic testing. A disease must be identifiable for a clinician to develop a candidate gene list. However, rare diseases often present atypically and, due to their progressive nature, patients may present to a clinician before recognizable symptoms have developed [19]. The majority of rare disease patients who receive a genetic diagnosis are diagnosed with their first genetic test, indicating diminishing returns to additional testing [19]. For many patients, the genetic etiology of their condition will remain elusive and there will be no satisfactory end to their diagnostic odyssey.

4.2.2 Opportunity for genomics

Genomic testing presents an opportunity to expedite the diagnostic testing process for rare diseases, enabling access to earlier interventions where available. In contrast to single-gene testing, genomic approaches are comprehensive and minimize the need for a candidate genes list prior to testing (Fig. 4.2). Genomic technologies can simultaneously interrogate all chromosomes (CMA), multiple genes (multi-gene panels), all protein coding regions of genes (whole exome sequencing; WES) or the whole genome (whole genome sequencing; WGS) [19]. Early-stage evidence suggests genomics provides increased diagnostic yield when used in conjunction with, or instead of, traditional testing methods [19].

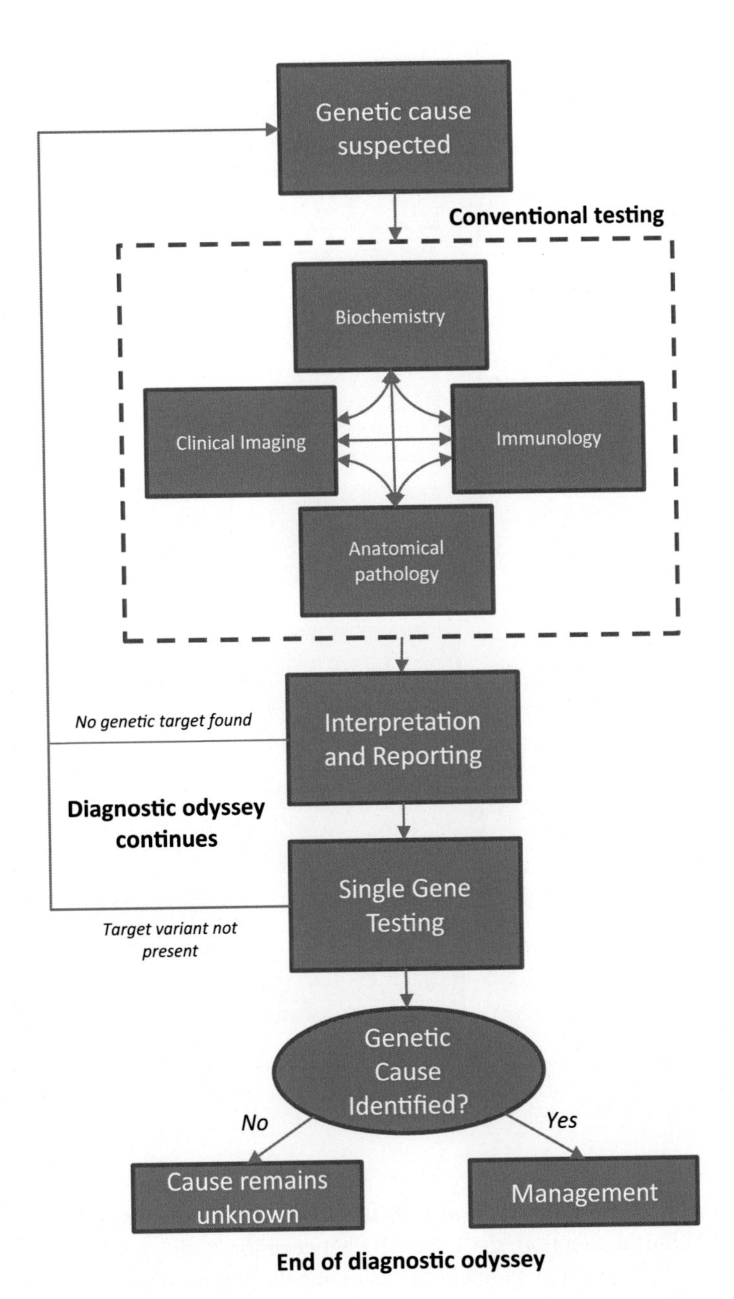

FIG. 4.1

Traditional diagnostic odyssey.

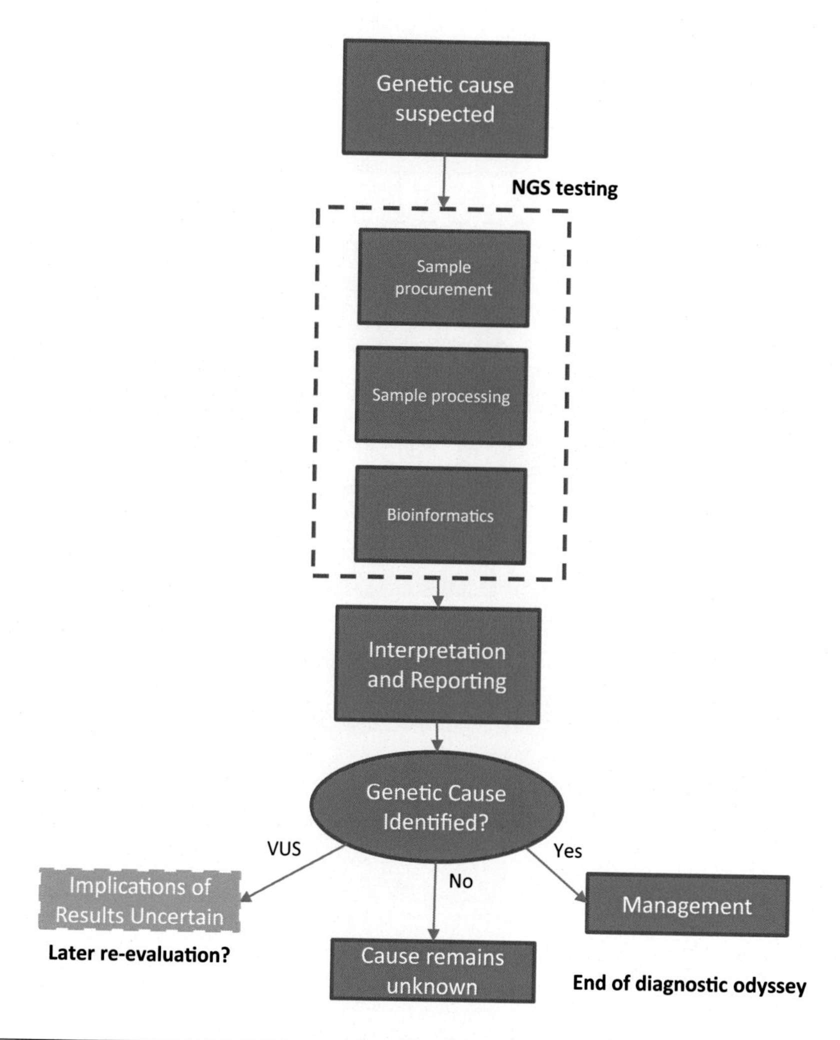

FIG. 4.2

NGS diagnostic odyssey.

4.2.2.1 Clinical and healthcare system implications of genomic testing

Prior to widespread clinical adoption, the trade-offs between the costs and benefits of genomic tests must be understood [20]. A recent meta-analysis examining the diagnostic utility of WES/WGS in children with suspected genetic conditions found mean diagnostic yields of 42% for WGS and 36% for WES [21]. This was contrasted against a mean diagnostic yield of 10% for CMA. The analysis reported clinical utility (i.e., diagnosis associated with treatment change) was 27% for WGS, 17% for WES, and 6% for CMA. Yet, improved diagnostic yield must be evaluated within the context of informational complexity and uncertainty and the value patient and families put on such information.

Single-gene tests and CMA are increasingly being replaced by multi-gene panels or WES/WGS, which reveal highly complex and often uncertain information. An example is the detection of genetic variants that confer unknown disease pathogenicity, termed "variants of unknown significance" (VUS) [22,23]. Increased VUS detection can facilitate efforts to reclassify such variants as either pathogenic or benign. To date, the majority of VUS are reclassified as benign, but this can take years to establish [24–26]. In general, it is recommended that VUS are not used to guide clinical decisions because their impact on disease risk is unknown and may never be determined [22,27,28]. Despite this recommendation, VUS are commonly identified by laboratories and increasingly returned to patients [29]. Debate exists regarding the value of returning VUS to patients on the basis that doing so may complicate medical decision-making, induce patient anxiety, and promote the uptake of potentially unnecessary medical interventions. These potentially harmful downstream consequences result from both patient and clinician misinterpretation of the pathogenic ambiguity attributable to VUS [27,30]. Rare diseases pose a major challenge to this effort because evidence informing pathogenic validation is limited [31].

In addition to genetic findings with uncertain pathogenicity, testing may also provide clinicians and patients with information about pathogenic variants unrelated to the initial reason for testing, termed secondary findings (SFs) [32]. In some cases SFs are clinically actionable, and effective prevention or medical interventions are available for the identified disorders. Alternatively, SFs may provide information about pathogenic variants for which no clinical action may be taken. Recommendations do exist regarding the return of SFs, however, there is ongoing debate about whether, and under what circumstances, they should be returned to patients [32,33]. Although the issues of informational complexity and uncertainty are not unique to the rare disease context, they are especially pertinent when considering the frequent lack of available treatment and clinical management options.

4.3 Economic evaluations of genomics in rare diseases

Economic evaluation facilitates comparison between the incremental expenditures of a given course of action and its incremental benefits. Published economic evaluations in pediatric rare disease settings often report that NGS is more effective than traditional testing pathways when establishing diagnoses, suggesting their potential to reduce the length of the diagnostic odyssey and mitigate its associated costs. The literature is not aligned on whether the application of NGS delivers these potential cost-savings, however. For example, the incremental cost-effectiveness of NGS versus CMA has been reported to be between CDN $25,458 and CDN $46,424 per additional diagnosis for autism spectrum disorder (ASD) [34], between US$2045 and US$6327 per additional diagnosis for suspected monogenic diseases when NGS is second line and last resort [35], respectively, and up to AUS$31,144 per quality-adjusted life year gained (QALY) for children with monogenic disorders [36]. When NGS is used as a first-line test, Stark et al. suggest WES is cost saving. Similarly, Yeung et al. report that first-line singleton WES results in a 42% diagnostic yield compared to 23% for traditional testing in children with complex monogenic disorders, with cost savings of US$2583 [37].

The cost-effectiveness of NGS in rare diseases will be impacted by the type of genomic technology (e.g., WGS/WES, multi-gene panel), the potential to mitigate the diagnostic odyssey, and the clinical and patient utility provided. Understanding and characterizing these impacts is crucial when conducting economic evaluations.

The process of conducting an economic evaluation can be summarized by four steps:

(1) Determine the new or existing health technology being evaluated, the relevant comparator, appropriate time horizon, and the study design for evaluating differences.
(2) Estimate the health, non-health, and process outcomes associated with using these competing technologies [38].
(3) Estimate the costs associated with using these competing technologies from the perspective of the relevant stakeholder.
(4) Describe the trade-offs between incremental costs and incremental outcomes for competing technologies, subject to uncertainty.

In the following section, we discuss these steps, their associated challenges, and examples of solutions proposed in the literature. We then summarize recommendations for future analyses and highlight remaining evidentiary gaps.

4.3.1 Step 1: Determine comparators and study design

When conducting an economic evaluation, researchers must answer two important questions. First, what new or existing policy, technology, or service is being evaluated? Second, what is this technology being compared to? Table 4.1 contains some examples of technologies, comparators, data sources, and study designs researchers may consider when designing their evaluations of genomics in rare diseases (Fig. 4.3).

Economic evaluations in rare diseases face challenges when identifying comparators. When used for diagnosis, genomic approaches are often adopted *in conjunction with* conventional approaches [19]. The relevant policy question typically asks *when* along the diagnostic pathway genomic tests should be introduced, rather than *if*. Randomized controlled trials (RCTs) are uncommon for evaluating genomic applications, as sample sizes required to account for genomic-level heterogeneity are often prohibitively large. In the absence of RCT data, counterfactuals for genomic approaches are typically unknown necessitating the use of observational data for conducting comparative evaluations. Such observational studies may suffer from selection bias unless appropriate measures are taken to account

Table 4.1 Step 1. Potential interventions, comparators, and data sources for economic evaluations of NGS in rare diseases.

Intervention	Comparator	Study Design
• Adopting a novel form of genomics • Moving genomics earlier in the diagnostic pathway	• Diagnostic standard of practice with conventional methods • Using genomics as a second-line diagnostic technique	• A randomized study cohort • Administrative record analysis with matched controls
	• Diagnostic standard of practice with another NGS method (e.g., a targeted panel array) • Multiple novel genomic comparisons	• Patients as their own historical controls • Modeled or hypothetical cohort based on assumptions grounded in literature

FIG. 4.3

Step 1.

for observed and unobserved confounding. Selection bias occurs when: (1) there is a systematic difference between those selected for the study and the target population for the technology and/or (2) there are systematic differences between comparison groups experiencing the intervention versus standard care. Selection bias can lead to erroneous conclusions around clinical and cost-effectiveness.

Published studies are beginning to apply quasi-experimental methods to mitigate selection bias while evaluating different diagnostic pathways for genomics in rare diseases. Quasi-experimental methodologies aim to mimic the conditions of a RCT. For example, these approaches can allow patients undergoing genomic testing to serve as their own historical controls or match selected patients with eligible controls. The following section summarizes selected examples from the literature where quasi-experimental methods were used to facilitate economic evaluations of genomics in rare diseases.

4.3.1.1 Selected literature examples

Using a trial cohort to compare testing methods—Sagoo et al. (2015)

Sagoo et al. [39] analyzed a retrospective cohort of 1590 patients in the United Kingdom seeking a diagnosis for learning disorders with a suspected genetic etiology. The authors pooled data from a previously collected study cohort, where participants had received various first-line tests for diagnosing patients with learning disorders. One group was diagnosed using first-line CMA, while the second cohort was idiopathic and underwent CMA as second line method.

In this study, the authors pooled existing data for two distinct patient groups that had visited the same laboratory facilities. This allowed them to estimate the cost-effectiveness of first- vs. second-line use of CMA, using a 1-year time horizon. The authors were able to capitalize on a pre-existing research study that had been conducted during the same time period as a large-scale policy change, which resulted in a natural experiment. Researchers seeking to establish their comparator groups by replicating Sagoo et al.'s approach would need access to a similar dataset. Additional care must be taken to ensure that the groups undergoing comparison are either similar or that covariate adjustment is used to control for confounding; failure to do so will introduce bias into the study.

Using a retrospective cohort to estimate impact of testing sequence—Schofield et al. (2017)

A study conducted by Schofield et al. [40] examined a cohort of 56 Australian pediatric patients diagnosed with neuromuscular diseases. Patients underwent first-line traditional diagnostic testing and, if a diagnosis was not achieved (n=30), subsequently received WES.[a] In this study, the authors compared two different forms of genomic testing against standard practice to evaluate multiple scenarios for technology adoption.

The authors capitalized on the availability of retrospective data for a well-characterized cohort that had undergone multiple tiers of diagnostic testing. This approach enabled the estimation of what would have happened had those patients received first-line genomic testing involving WES, rather than standard diagnostic testing. This study was limited by a relatively small sample size of 30 cases from a single institution. The study also did not consider any health care resource utilization beyond diagnosis. Further, the time horizon is not clearly reported in the evaluation, and the authors do not consider healthcare resource utilization beyond diagnosis. As a result, the authors are unable to estimate the downstream patient and health system impact of WES.

Comparing two different genomic approaches in a retrospective cohort—Hayeems et al. (2017)

A study conducted in a Canadian pediatric hospital by Hayeems et al. [41] compared two different genomic tests, WGS and CMA, for difficult-to-diagnose pediatric disorders. The study cohort included 101 children who had had an idiopathic disorder after standard testing. If a diagnosis was still not achieved after CMA (n=93) they went on to receive WGS. The authors examined clinical healthcare utilization resulting from these two different diagnostic approaches. Downstream impacts were evaluated for a period of one year following NGS testing.

This study is another example wherein a group's diagnostic history was used as a comparator to estimate the impacts of adopting a new technology. A key difference between this study and Schofeld et al. is the relevant policy question, which centered on two different genomic technologies. Despite this difference, the authors still used standard clinical practice (CMA, in this case) as the most relevant comparator.

[a]Three patients were diagnosed with a multi-gene panel that tested for neuromuscular diseases.

4.3.1.2 Recommendations

The cited studies illustrate how researchers are beginning to address the challenges of comparative evaluation of testing strategies in rare diseases. Study designs can take advantage of existing cohorts where there is a difference in diagnostic practice, but care must be taken to ensure comparability between cohorts to mitigate the potential for bias. When an existing cohort is not available, researchers can consider using patients as their own historical controls, comparing pre-genomic to post-genomic diagnostic processes and outcomes. To make results relevant for decision-making, researchers should assess change from the current standard of practice rather than compare new technologies to each other. If multiple new technologies are being considered, it may be appropriate to compare them to standard practice and each other, through secondary scenario analysis.

All studies described specified diagnosis as their primary endpoint and did not consider future health or non-health outcomes that patients or families may value. The application of short time horizons fails to capture downstream outcomes attributable to NGS technologies, beyond diagnostic yield. While measuring these outcomes is challenging, they are a crucial component of the economic evaluation of genomic technologies.

With this in mind, economic evaluations of genomics in rare diseases should:

- Use the current standard of practice as the most relevant comparator.
- Include multiple comparators to inform scenario analysis, as appropriate.
- In the absence of RCT data, apply quasi-experimental methods to mitigate selection bias.
- Specify a study time horizon that is appropriate for measuring relevant costs and effects, through consultation with clinical experts.

4.3.2 Step 2: Estimate health and non-health outcomes

After finalizing competing technologies and study design, researchers must characterize the value of expected outcomes of each technology, including health, non-health (e.g., resolving uncertainty, disease stigma), and process outcomes (e.g., time waiting for results, who delivers testing). Depending on the rare disease, genomics-driven diagnosis can impact individual patients, healthcare systems, and the broader population. Table 4.2 provides examples of outcomes that may result from a change in diagnostic approach (Fig. 4.4).

Economic evaluations frequently consider technologies that improve duration and/or quality of life [42,43]. In these evaluations, impacts can be expressed in terms of survival (e.g., life years gained;

Table 4.2 Step 2. Potential outcomes to measure in economic evaluations of NGS in rare diseases.

Diagnostic	Medical	Other
• Number/proportion of patient cohort that achieve a diagnosis • Time to achieve a diagnosis • Number/proportion of relatives testing positive for variant	• Number/proportion receiving a change in management • Number of adverse effects avoided • Inappropriate treatments avoided	• Traditional health outcomes: level of disability, quality of life, survival, quality adjusted survival • Behavior change (e.g., using in vitro reproductive techniques among carriers) • Personal utility

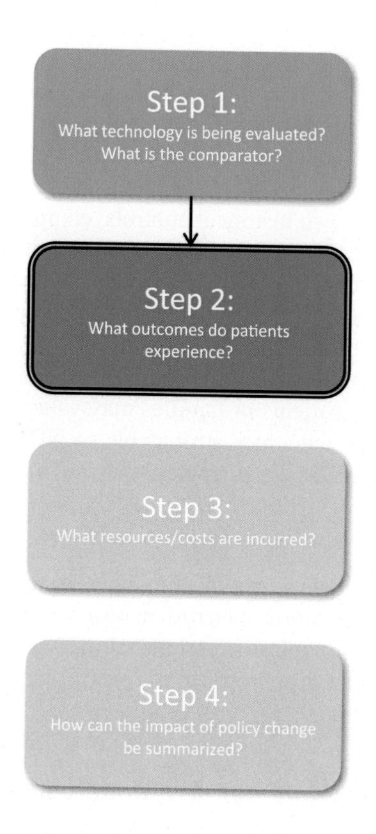

FIG. 4.4

Step 2.

LYG) or quality-adjusted survival (i.e., QALYs). The QALY is a measure that encompasses the amount of time a person spends in a state of health, and the preference-based health benefit, termed utility, from being in that state [42,43].

Using typical measures to assess benefits is a challenge in rare diseases contexts. For many rare diseases, particularly developmental and intellectual conditions, diagnosis will not necessarily improve survival. If diagnosis leads to a change in clinical management, these effects are difficult to quantify using the QALY framework. Ethical concerns are raised when researchers assume different utilities for a person with an intellectual disability versus someone neurotypical [44–46].

To estimate outcomes for genomics in rare diseases, researchers have used a variety of endpoints that include and transcend conventional measures of survival or quality-adjusted survival. The following section describes a selection of these endpoints.

4.3.2.1 Selected literature examples

Cost per additional diagnosis—Vissers et al. (2017)

A Dutch study conducted by Vissers et al. [47] examined a prospective cohort of 150 patients referred to a pediatric neurology center for evaluation. Patients were evaluated with two different diagnostic pathways in parallel, allowing for direct comparison of diagnostic yield between conventional methods (magnetic resonance imaging (MRI) and targeted single-gene tests) and WES. The authors considered outcomes in terms of diagnostic yield. While changes in therapeutic management were briefly described for some patients, the corresponding resource utilization and outcomes were not.

Cost per additional diagnosis and changes in management—Soden et al. (2014)

In a retrospective cohort study, Soden et al. [48] estimated the cost-effectiveness of using genomic testing(WES and WGS) to diagnose 119 American children with a suspected monogenetic cause to their neurodevelopmental disorders. Patients served as their own historical controls when estimating costs and outcomes of genomic testing compared to standard diagnostic testing. The study's primary outcome measured the number of additional diagnoses achieved through genomic testing. The authors also reported the clinical consequences that the 49% of patients with new diagnoses experienced. These consequences included changes in patient management, new drug or dietary treatments, discontinuation of ineffective treatments, and evaluations for possible disease complications that would not have otherwise been undertaken.

Cases avoided through testing—Azimi et al. (2016)

Azimi et al. [49] used decision analytic modeling to estimate costs and outcomes of either a multi-gene panel or conventional carrier testing for 14 genetic diseases in a simulated population of 1,000,000 couples. The model had three arms: carrier testing before conception, carrier testing after pregnancy, and no carrier testing. Primary outcomes included: birth of a child with a genetic disorder, birth of a child without a genetic disorder, and instances where couples adjusted their reproductive plans to avoid having a child with a disorder. This category included choosing to adopt, terminating a pregnancy, or using a gamete donor, among other interventions. The authors used a combination of literature-based values and simplifying assumptions to estimate the incremental survival impact of carrier testing. In this study, QALYs were estimated for *relatives* of probands, but not the probands themselves (Box 4.1).

How is personal utility estimated?

Researchers use stated preference methods, typically discrete choice experiments (DCEs), to quantify personal utility [38]. DCEs are based on economic utility theory and ask respondents to repeatedly choose between two or more competing alternatives in a series of questions, called choice tasks. Within each choice task, alternatives are described by their characteristics, called attributes. Attributes vary across a range of levels affecting the value of each alternative and may relate to either health or non-health outcomes. Individuals' responses in choice tasks reflect trade-offs between different attributes and can be used to derive utility estimates.

There is evidence that patients value health and non-health outcomes related to genomics in rare diseases. For example, Regier et al. [55] used a DCE to estimate parents' personal utility for genomic testing to diagnose idiopathic developmental disability (IDD). Questionnaires were administered to 89 parents of children with IDD. Results indicated that parents valued health, non-health, and process attributes related to genomic testing. Attributes included diagnostic yield, time spent waiting for results, and the amount participants would pay out of pocket for testing. By including the "cost to you" attribute, the authors were able to estimate willingness to pay (WTP) for an additional diagnosis (Fig. 4.5).

BOX 4.1 Genomics and personal utility

Rare diseases with no effective treatments available raise an important question: is there value in a diagnosis itself? In particular, do people value genomic information even if it doesn't change clinical outcomes? If so, how do we estimate this value and should we incorporate it into economic evaluation?

What is personal utility?

Personal utility relates to the value that individuals ascribe to genomic information irrespective of its ability to improve health status [38,50]. This value can reflect benefits and harms from a rare disease diagnosis as well as from genomic information itself. Benefits can include increasing knowledge about one's self and one's disease, informing preparation and planning for the future, and providing potentially useful information to one's family [51]. Diagnosis may also cause harm, as patients may worry about being stigmatized or losing social support or privacy. Each of these non-health outcomes can impact a person's wellbeing, indicated by positive or *negative* personal utility, termed disutility, but are not captured in a QALY framework.

In addition to patients and their families, it may also be important to measure personal utility for genomic testing among members of the general public. In socialized healthcare systems, society invests in technologies, and it can be relevant to understand what value members of the public place on these kinds of non-health outcomes, and whether they differ from that of patients [52,53].

Understanding personal utility may be crucial to ensuring that genomic tests are delivered appropriately when diagnosing rare diseases. Personal utility can guide policies around whether it is appropriate to disclose non-actionable genetic findings.

When should personal utility be included in economic evaluations?

There is considerable debate in the literature regarding the inclusion of personal utility in economic evaluations. Recent recommendations for economic evaluation support a reference case that accounts for health and non-health outcomes [54].

Information on personal utility may be especially important when conventional analyses focusing exclusively on health outcomes do not provide clear recommendations about cost-effectiveness. These circumstances might arise when willingness to pay (WTP) for a particular outcome, such as an additional diagnosis, is unknown. WTP thresholds can be approximated using personal utility estimates. Personal utility can also be informative when mean incremental cost-effectiveness ratios lie close to established WTP thresholds. In these situations, alternatives under consideration may seem equivalent without the inclusion of personal utility.

Would you prefer postnatal test A, postnatal test B or neither?

	Test A	Test B	Neither Test
Number of children tested whose genetic condition is identified with this test	10 children in 100 with DD who are tested	14 children in 100 with DD who are tested	In this scenario, you would prefer neither of the tests to be conducted
Time waiting for the results of the test	1 week	3 weeks	
Cost to you	$750	$1100	

In this situation would you choose (please select only one of the following options):

___ 1) Test A

___ 2) Test B

___ 3) Neither

FIG. 4.5

DCE questionnaire.

4.3.2.2 Recommendations

The studies described have proposed methods for estimating the impact of genomic testing on health and non-health outcomes. When appropriate, conventional methods of measuring survival and/or QALYs were occasionally used. It was more typical, however, for exercises to use rate of diagnosis as their endpoint. Using such an endpoint does not capture the value or benefit of all the relevant health and non-health consequences of a diagnosis and is inappropriate for a full economic evaluation.

Much of the research in rare diseases occurs in a pediatric context. Researchers note a number of methodological challenges when it comes to using preference-based tools to evaluate children's health [56]. Over the past decade, researchers have developed new approaches to respond to these challenges, including childhood-specific valuation methods [57]. Use of these tools has increased over time [58]. Challenges remain when it comes to adapting these methods for intellectual disability, which is a common reason for children to undergo genomic testing.

Being mindful of these changes, researchers conducting economic evaluations of genomics in rare diseases should:

- Use index measures such as the QALY if possible, to allow comparisons to other economic analyses using QALYs.
- Measure and/or incorporate preference-based personal utility into studies.
- Report diagnostic yield when it is either not possible or inappropriate to use QALYs or other preference-based measures.
- Consider reporting non-diagnostic outcomes, such as disease event rates and diagnostic odyssey length, in sensitivity analyses.

4.3.3 Step 3: Estimate costs

The adoption of any new technology is usually accompanied by increased expenditure. Genomic approaches require sophisticated laboratory equipment and highly specialized labor to conduct tests and interpret results. Table 4.3 lists some of the costs that may occur when applying genomic testing to diagnose rare diseases (Fig. 4.6).

Table 4.3 Step 3. Potential resource utilization to measure in economic evaluations of NGS in rare diseases.

Diagnostic	Medical	Other
• Sample collection, processing, and bioinformatics for genomic testing	• Treatments (e.g., drugs, surgery, other therapies)	• Patient out-of-pocket expenses (e.g., travel to appointments, special diets, other uninsured services) • Changes in economic participation, including productivity losses due to caregiving
• Conventional diagnostic tests: imaging, biochemistry, pathology • Supplies, overheads, and equipment • Confirmatory testing	• Follow-up or surveillance appointments • Genetic counseling	

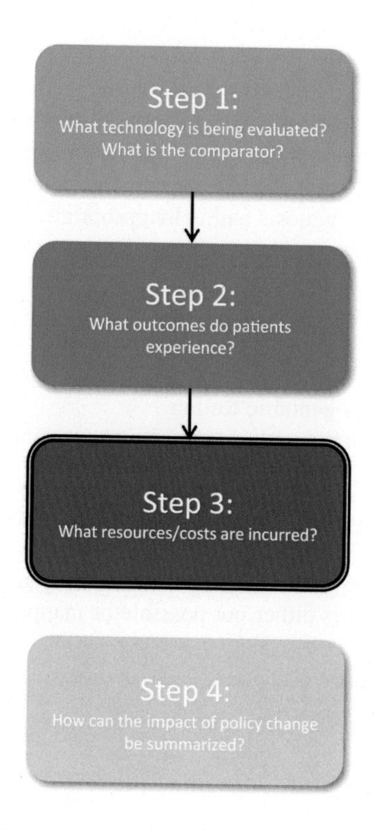

FIG. 4.6

Step 3.

Despite these additional testing costs, genomic technologies in some contexts may reduce resource utilization through reductions in the length of the diagnostic odyssey and avoidance of future exploratory testing. Additional downstream impacts of diagnosis include changes in clinical management, adverse events avoided, and discontinuation of ineffective treatments. These impacts may drive the results of an economic evaluation but will be ignored in studies whose endpoint is diagnosis.

Researchers must decide which costs to include in their analyses. Failing to describe resource inputs for genomic testing and corresponding cost estimation can make accurate comparisons across studies difficult, even when genomic technologies being evaluated are ostensibly the same.

Studies in the literature have used a variety of methods to estimate the costs of genomics-guided diagnosis, including surveys, microcosting exercises, electronic medical record (EMR) reviews, and costing models. We discuss selected examples of these methods in the following section.

4.3.3.1 Selected literature examples

Costing surveys for laboratory tests—Wordsworth et al. (2007)

A cost-effectiveness study conducted by Wordsworth et al. [59] estimated the costs of CMA compared to conventional karyotyping for diagnosing IDD using costing surveys. The authors administered surveys in four National Health Service (NHS) hospitals across the United Kingdom. Surveys collected information on staff time, consumables, and capital required to test a typical tissue sample from time of sample collection to reporting of results. Staff time was converted to costs using NHS salary rates. Equipment costs were determined from purchase prices, with capital costs estimated based on the lifetime depreciated cost of the item. The scope of the study was limited to costs occurring within the diagnostic pipeline. As a result, the authors did not consider costs occurring beyond diagnosis.

Bottom-up microcosting of clinical activity—Tsiplova et al. (2017)

Tsiplova et al. [34] estimated the costs of a range of genomic approaches, including WES, WGS, and CMA, to diagnose ASD. The authors focused on a pediatric hospital in Canada and estimated costs of genomic testing, interpretation of results, and subsequent visits to clinic over five years. Cost components included:[b]

- labor for sample logistics and data processing
- data storage and computation tasks
- equipment purchasing and maintenance (including labor)
- overheads

Laboratory workflow procedures from 155 separate genomic testing activities informed the cost calculations. Labor was calculated by multiplying salaries with the time involved for each workflow procedure. Equipment costs were calculated based on purchase prices, expressed on a per-test basis. While this study provided a comprehensive description and estimated value for each cost component, it did not consider downstream impacts, such as genomics-informed treatment costs. The exercise also excluded the costs of bioinformatics software costs, and resource utilization associated with validation, quality control, and updating/testing the diagnostic pipeline.

4.3.3.2 Recommendations

The studies described used a variety of methods to estimate the costs of diagnosing rare diseases. Among these studies, and in the literature more generally, there is a lack of consistency in which types of healthcare resources are included when costing a diagnosis. The studies varied in whether or not resources beyond diagnosis were considered.

To date, the majority of costing analyses for genomic technologies, particularly WGS, focus on research settings [60]. As genomics becomes more widely integrated, jurisdictions may find ways to consolidate genomic testing activities into centralized labs with higher throughput. Researchers conducting economic evaluations of genomic technologies should bear this in mind, as cost estimates published from laboratory sources may overestimate costs of widespread use.

With these points in mind, economic evaluations of genomics in rare diseases should:

- Ensure that the costs of genomic technologies reflect all necessary components, including labor, equipment, capital, informatics and interpretation, counseling, and return of results.
- Pay attention to the study perspective and include costs for all relevant system events, not merely those involved for diagnosis.

[b]There is a great deal of variability in the published literature when it comes to fully describing resource utilization in genomic testing. The paper by Tsiplova et al. is notable in the level of detail it provides about what they included, what they didn't include, and how they derived the associated costs. This paper may serve as a useful template for other researchers.

- When available, use patient charts and administrative health records to identify resource inputs. In the absence of this information, activity- or survey-based bottom-up microcosting should be conducted.
- Regardless of approach, always provide detailed descriptions of resource inputs and corresponding unit cost estimates for comparability.

4.3.4 Step 4: Estimate relationship between costs and outcomes

Economic evaluations allow decision-makers to determine if an intervention offers value for money. If a genomic technology is funded, how will costs and outcomes *change*? If multiple genomic technologies are available, which approach is efficient given a valued outcome? (Fig. 4.7).

Economic evaluations typically summarize value for money in a single measure: the incremental cost-effectiveness ratio (ICER). The ICER is the expected change in costs divided by the expected change in outcomes and is commonly expressed in terms of the incremental cost per QALY gained. Decisions are made by comparing the ICER to a threshold value, termed λ, representing a decision-maker's willingness to pay for a unit of health improvement. Given that QALYs may not be sensitive to changes in diagnosis for rare diseases, these conventional metrics may not be applicable.

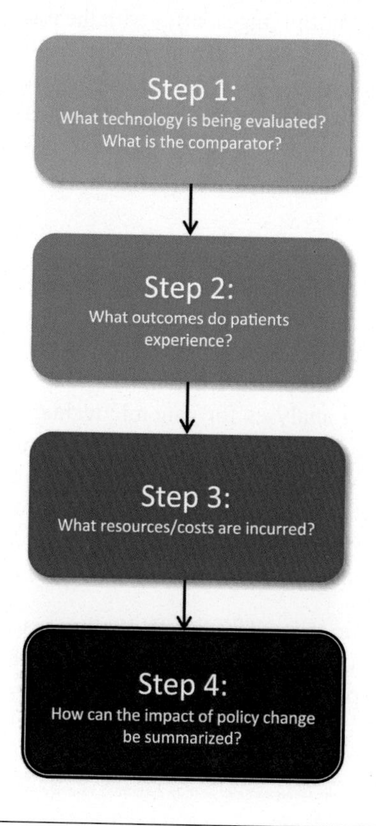

FIG. 4.7

Step 4.

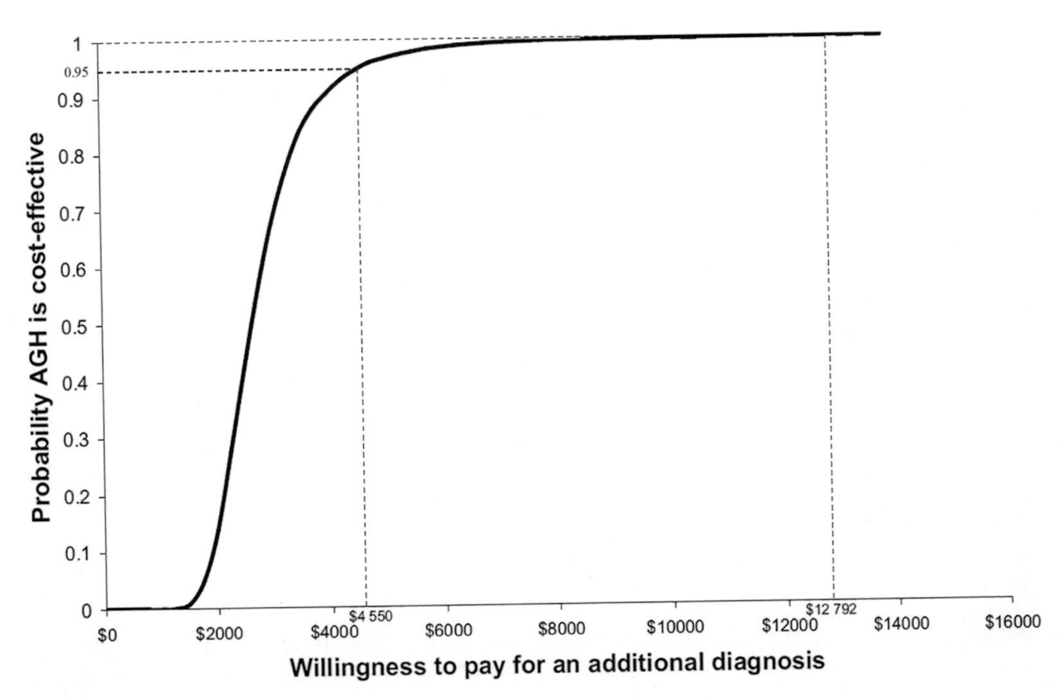

FIG. 4.8

Cost-effectiveness acceptability curve (CEAC). Figure from Regier et al. [61] examining the cost-effectiveness of CMA versus conventional testing (employing karyotyping as first line). The decision uncertainty surrounding the cost-effectiveness of CMA is examined using the CEAC. The CEAC plots the probability that CMA will be cost-effective at different thresholds of WTP that decision-makers may have for an effectiveness gain. This figure reports a 95% probability the CMA strategy will be cost-effective if decision-makers are willing to pay $4550 for an additional diagnosis. Regier [62] suggests that parents' WTP is $12,792 per additional diagnosis. The probability that CMA will be cost-effective at this WTP threshold is greater than 99%.

In rare diseases, estimates of incremental costs and effects are subject to considerable decision uncertainty—the probability of making the wrong decision—due to incomplete information resulting from statistical imprecision. Economic analyses must quantify the impacts of imprecision on the expected relationship between costs and effects, that is, how confident a decision-maker can be about whether adopting a new technology is cost-effective. Box 4.2 discusses this uncertainty and how it affects research design in greater detail.

4.3.4.1 Selected literature examples

Cost-effectiveness, net benefit, and personal utility—Regier et al. (2010)

Regier et al. [61] constructed a decision analytic model to investigate the cost-effectiveness of CMA to diagnose children with IDD. Diagnosis using conventional testing methods was the comparator. The first-line genetic test in the conventional testing pathway was karyotyping. The probability of receiving a genetic diagnosis with a karyotype differed between those with and without suspected trisomies 21, 18, or 13. If a karyotype did not provide a diagnosis, it was assumed that fluorescence in situ

BOX 4.2 Uncertainty in genomic diagnosis of rare diseases

Decision uncertainty

The technological landscape for genomic testing is changing rapidly and research on gene pathogenicity continues to evolve. This influx of information can increase decision uncertainty, as the cost-effectiveness of a genomic approach can change dramatically after a study's publication. Sources of uncertainty may include:

Costs	Effects
Sample processing time (labor cost)	Variant penetrance in target population
Staff salaries	Length of diagnostic odyssey
Treatments received in hospital	Length of hospital stay
Number of conventional diagnostic tests	Change in survival/QALYs

To ensure that the results of economic evaluations remain relevant for decision-makers, researchers must account for decision uncertainty arising from technological change. The impact of this uncertainty is particularly relevant in cases where decision-makers' WTP for an outcome of interest (e.g., change in diagnostic yield) is unknown or likely to differ across jurisdictions.

Probabilistic analysis is a method of quantifying overall decision uncertainty as a product of the uncertainty around each parameter in the evaluation. Cost-effectiveness acceptability curves (CEACs), illustrated in Fig. 4.8, are a useful graphical tool for presenting the probability that a technology will be cost-effective at various WTP thresholds.

Structural uncertainty and scenario analysis

Researchers may also consider the use of scenario analysis to estimate the impact that systemic and structural changes might have on cost-effectiveness. Additionally, scenario analysis can be used to explore potential policy changes beyond the observed evidence. The uncertainty around these changes, and the impact they have on decision-making, is termed "structural uncertainty." Scenario analysis uses observed values for costs and effects to examine a "what if?" circumstance: What if the cost of genomic testing decreased by $X? What if genomic testing was used earlier (or later) in the diagnostic process? What if an alternative policy was adopted that changed one or more service pathways? By changing a single variable (univariate) or several variables (multivariate), scenario analysis allows researchers to consider a broad range of potential policy changes, and how they might play out under different circumstances.

hybridization (FISH) would be conducted. For CMA, a karyotype would be performed for those with suspected trisomy disorders. Array genomic hybridization (AGH) was subsequently undertaken if a diagnosis was not established via karyotyping, and targeted FISH (in the parents and child) and karyotyping (in the child) would be applied to establish the relevance of any imbalance detected by CMA. Those children in whom trisomy disorders were not suspected had AGH testing as first line, followed by targeted FISH (in the parents and child) and karyotyping (in the child) to confirm the AGH finding and establish whether it occurred de novo. The costs and diagnostic yield of the two testing strategies were evaluated using the model, estimated through a synthesis of literature-published values and a retrospective chart review.

ICERs were estimated in terms of incremental cost per additional diagnosis. The authors used probabilistic analysis to generate CEACs. This approach avoided assuming a particular WTP threshold. Instead, decision-makers could decide which approach was most cost-effective based on their own program needs and budgets. Scenario analyses explored the impacts of genomic testing in a variety of plausible policy environments.

To represent patient-centered values in the evaluation, the authors used a cost-benefit approach that incorporated families' preference-based utility (based on their willingness to pay) in the primary economic outcome, the net benefit of CMA. Net benefit was expressed as the incremental cost subtracted from incremental WTP for an additional diagnosis ($NMB = \Delta WTP_{Diagnosis} - \Delta Cost_{Diagnosis}$). Decision uncertainty was characterized by creating a probabilistic net-benefit curve. The net benefit of CMA testing versus conventional testing was CDN$835 (95% confidence interval $203–$1616), indicating a greater than 95% chance that CMA was cost-effective.

Cost per QALY among relatives of affected individuals—Perez et al. (2011)

Perez et al. [63] conducted Markov modeling to evaluate the impact of testing family members of people with long-QT syndrome (LQTS) using a multi-exome panel. This study investigated the cost-effectiveness of three different methods for providing early interventions in asymptomatic relatives of probands (the first tested individual in the family). Methods included exome sequencing with treatment for variant-positive cases, treatment of all family members with beta blockers, and watchful waiting with treatment after symptoms develop. The comparator was watchful waiting without treatment. Costs and QALYs were measured over a 60-year time horizon. Cost estimates included drugs, hospitalizations and clinic visits, implantable devices, and commercially available gene tests.

The interventions were expected to yield QALY changes for asymptomatic relatives of probands. Genomic testing was more costly than watchful waiting but resulted in a QALY increase that the authors considered cost-effective. The authors also reported rates of other outcomes, including drug and device use and sudden cardiac deaths, for simulated cohort members. To account for model and parameter uncertainty, the authors conducted univariate and multivariate scenario analyses. Probabilistic analysis was not conducted.

Cost per additional diagnosis achieved—Tan et al. (2017)

Tan et al. [64] estimated the impact of applying WES at different timepoints in the diagnostic process versus never applying WES for a prospective cohort of 44 children. These children had a suspected monogenetic condition and received WES as part of their standard diagnostic pathway. The evaluation examined WES either after standard diagnostic procedures, at the point of clinical genetics assessment, or at first presentation. The total number of diagnoses achieved was the study's primary outcome and cost estimates were based on chart review. The authors estimated changes in costs by assuming earlier diagnosis would forego any additional testing procedures.

This cost-effectiveness study reported an ICER in terms of incremental cost per additional diagnosis. The authors found that using WES at the time of initial presentation yielded a higher number of diagnoses at a net cost saving compared to using WES at clinical genetics assessment. Bootstrapping techniques were used to extrapolate the uncertainty around cost per additional diagnosis within each diagnostic pathway. Decision uncertainty was expressed using 95% confidence intervals and by plotting bootstrapped cost-per-diagnosis values on the cost-effectiveness plane.

Costs and consequences of technology adoption—Córdoba et al. (2018)

Córdoba et al. [65] estimated the costs and consequences of introducing WES in a prospective cohort of 40 Argentinian adult and pediatric patients with neurogenetic conditions. The comparator was traditional diagnostic testing. Primary outcomes included number of diagnoses, average length of diagnostic odyssey, and number of outpatient visits. Patient-level costs included prior laboratory procedures received before WES and WES costs, though few details were provided on costing methods.

The authors do not suggest a particular value of WTP for any of the outcomes in the study but do note that the use of WES was both time- and cost-saving compared to standard practice. The authors note that few studies have evaluated the use of NGS technologies in middle-income countries. Since costs and specific resources are likely to differ widely between those countries, the authors chose to publish resource utilization and outcomes information separately rather than a single summary cost-effectiveness measure (i.e., a cost-consequence study).

4.3.4.2 Recommendations

Cost-effectiveness analysis is a common approach for evaluating the economic impacts of genomic testing in rare diseases. The studies described used a variety of methods to adapt the cost-effectiveness framework to a rare disease context. Few studies directly incorporated QALYs, as this measure would not be sensitive to meaningful differences in benefits. In these situations, ICERs were expressed as the incremental cost per additional diagnosis or through cost-benefit analysis using net benefit. Scenario analysis and probabilistic analysis were used to explore the impact of various forms of decision uncertainty.

With these points in mind, economic evaluations of genomic testing in rare diseases should:

- Use cost-effectiveness analysis, with incremental cost per QALY used as a reference case (i.e., the default method) where possible. Expressing cost-effectiveness as net benefit is preferred when incremental QALYs are near zero.
- In accordance with recommendations from the Second Panel on Cost-Effectiveness [54], use cost-benefit analysis when QALYs cannot be calculated or are insensitive to benefit In this context, benefit is measured in patients', families', or societal WTP for a valued outcome.
- Cost-consequence analysis may be particularly useful for making comparisons between healthcare jurisdictions, in circumstances where health economic evidence is sparse.
- Cost per diagnosis (i.e., a non-incremental measure) should not be used.
- If multiple technologies or diagnostic pathways are considered, use probabilistic analysis to generate CEACs with multiple WTP values. Use scenario analyses to express how potential technological changes might impact cost-effectiveness. Scenario analysis should also be used to examine structural uncertainty, especially in studies using decision models.

4.4 Conclusion

Economic evaluations provide decision-makers with important evidence about how best to allocate resources. Rare diseases pose unique challenges for economic evaluation. In this chapter, we have reviewed these challenges and highlighted ways that researchers are going about addressing them.

Important considerations when planning an economic evaluation include:

- What is the standard of practice when diagnosing a rare disease?
- What are the relevant outcomes that might change if genomic testing is adopted? Will it affect the survival of patients? Their family members?

- What data are available for measuring changes in diagnostic yield, costs, and other outcomes that genomic testing might produce?
- What are the effects on patients, and on the healthcare system, if a diagnosis is achieved? How long do these changes last, and on what timescale do they occur?
- How can all this information best be relayed to decision-makers?

These questions need to be addressed in the research design stage, with input from and in collaboration with clinicians and other healthcare professionals. The design should also include input from patients, who will have important information about the impact that a diagnosis has on their lives, particularly in terms of out-of-pocket expenses and changes in disease management. The design process should also incorporate patient and family preferences and its associated utility, particularly in the face of the complex information that comes with genomic testing (e.g., SFs, VUS) and in context to the value of diagnostic information (i.e., personal utility).

As genomic technologies become more widespread, some questions will require the development of new methods that combine theoretical paradigms from economics, medicine, psychology, and genomics. Pairing these paradigms with qualitative and quantitative data collection will be critical to supporting the appropriate adoption of genomics for solving the diagnostic odyssey in rare diseases. The challenge is great, but so may be the rewards in the form of better and more cost-effective health care with faster and more precise diagnoses.

Acknowledgment

This research was funded by Genome British Columbia (GEN001).

References

[1] Canadian Organization for Rare Disorders. CORD patient survey overview, 2015. Available from: www.raredisorders.ca/our-work.

[2] Drummond MF, Wilson DA, Kanavos P, Ubel P, Rovira J. Assessing the economic challenges posed by orphan drugs. Int J Technol Assess Health Care 2007;23(1):36–42.

[3] O'Sullivan BP, Orenstein DM, Milla CE. Pricing for orphan drugs: will the market bear what society cannot? JAMA 2013;310(13):1343–4.

[4] European Organisation for Rare Diseases. Rare diseases: understanding this public health priority: Eurodis; 2005.

[5] Basel D, McCarrier J. Ending a diagnostic odyssey: family education, counseling, and response to eventual diagnosis. Pediatr Clin N Am 2017;64(1):265–72.

[6] Molster C, Urwin D, Di Pietro L, Fookes M, Petrie D, van der Laan S, et al. Survey of healthcare experiences of Australian adults living with rare diseases. Orphanet J Rare Dis 2016;11:30.

[7] Nunn R. "It's not all in my head!"—the complex relationship between rare diseases and mental health problems. Orphanet J Rare Dis 2017;12(1):29.

[8] Lohmann K, Klein C. Next generation sequencing and the future of genetic diagnosis. Neurotherapeutics 2014;11(4):699–707.

[9] Tripathy D, Harnden K, Blackwell K, Robson M. Next generation sequencing and tumor mutation profiling: are we ready for routine use in the oncology clinic? BMC Med 2014;12(1):140.

[10] Aronson JK. Rare diseases and orphan drugs. Br J Clin Pharmacol 2006;61(3):243–5.
[11] Dawkins HJS, Draghia-Akli R, Lasko P, Lau LPL, Jonker AH, Cutillo CM, et al. Progress in rare diseases research 2010-2016: an IRDiRC perspective. Clin Transl Sci 2018;11(1):11–20.
[12] Bavisetty S, Grody WW, Yazdani S. Emergence of pediatric rare diseases: review of present policies and opportunities for improvement. Rare Dis 2013;1, e23579.
[13] Evans WR, Rafi I. Rare diseases in general practice: recognising the zebras among the horses. Br J Gen Pract 2016;66(652):550–1.
[14] Yoon PW, Olney RS, Khoury MJ, Sappenfield WM, Chavez GF, Taylor D. Contribution of birth defects and genetic diseases to pediatric hospitalizations. A population-based study. Arch Pediatr Adolesc Med 1997;151 (11):1096–103.
[15] Dodge JA, Chigladze T, Donadieu J, Grossman Z, Ramos F, Serlicorni A, et al. The importance of rare diseases: from the gene to society. Arch Dis Child 2011;96(9):791–2.
[16] Schieppati A, Henter JI, Daina E, Aperia A. Why rare diseases are an important medical and social issue. Lancet (London, England) 2008;371(9629):2039–41.
[17] Luzzatto L, Hollak CE, Cox TM, Schieppati A, Licht C, Kaariainen H, et al. Rare diseases and effective treatments: are we delivering? Lancet (London, England) 2015;385(9970):750–2.
[18] Ouyang L, Grosse S, Raspa M, Bailey D. Employment impact and financial burden for families of children with fragile X syndrome: findings from the National Fragile X Survey. J Intellect Disabil Res 2010;54 (10):918–28.
[19] Sawyer SL, Hartley T, Dyment DA, Beaulieu CL, Schwartzentruber J, Smith A, et al. Utility of whole-exome sequencing for those near the end of the diagnostic odyssey: time to address gaps in care. Clin Genet 2016;89 (3):275–84.
[20] Davies SC. Annual report of the chief medical officer 2016: generation genome. London: Department of Health; 2017.
[21] Clark MM, Stark Z, Farnaes L, Tan TY, White SM, Dimmock D, et al. Meta-analysis of the diagnostic and clinical utility of genome and exome sequencing and chromosomal microarray in children with suspected genetic diseases. NPJ Genom Med 2018;3:16.
[22] Welsh JL, Hoskin TL, Day CN, Thomas AS, Cogswell JA, Couch FJ, et al. Clinical decision-making in patients with variant of uncertain significance in BRCA1 or BRCA2 genes. Ann Surg Oncol 2017;24(10):3067–72.
[23] Richards S, Aziz N, Bale S, Bick D, Das S, Gastier-Foster J, et al. Standards and guidelines for the interpretation of sequence variants: a joint consensus recommendation of the American College of Medical Genetics and Genomics and the Association for Molecular Pathology. Genet Med 2015;17(5):405.
[24] Murray ML, Cerrato F, Bennett RL, Jarvik GP. Follow-up of carriers of BRCA1 and BRCA2 variants of unknown significance: variant reclassification and surgical decisions. Genet Med 2011;13(12):998.
[25] Wright M, Menon V, Taylor L, Shashidharan M, Westercamp T, Ternent CA. Factors predicting reclassification of variants of unknown significance. Am J Surg 2018;216.
[26] Eccles B, Copson E, Maishman T, Abraham JE, Eccles D. Understanding of BRCA VUS genetic results by breast cancer specialists. BMC Cancer 2015;15(1):936.
[27] Pollard S, Sun S, Regier DA. Balancing uncertainty with patient autonomy in precision medicine. Nat Rev Genet 2019;20(5):251–2.
[28] Makhnoon S, Garrett LT, Burke W, Bowen DJ, Shirts BH. Experiences of patients seeking to participate in variant of uncertain significance reclassification research. J Commun Genet 2018;1–8.
[29] Wynn J, Lewis K, Amendola LM, Bernhardt BA, Biswas S, Joshi M, et al. Clinical providers' experiences with returning results from genomic sequencing: an interview study. BMC Med Genet 2018;11(1):45.
[30] Elliott AM, Friedman JM. The importance of genetic counselling in genome-wide sequencing. Nat Rev Genet 2018;1.

[31] Gainotti S, Mascalzoni D, Bros-Facer V, Petrini C, Floridia G, Roos M, et al. Meeting patients' right to the correct diagnosis: ongoing international initiatives on undiagnosed rare diseases and ethical and social issues. Int J Environ Res Public Health 2018;15(10):2072.

[32] Green RC, Berg JS, Grody WW, Kalia SS, Korf BR, Martin CL, et al. ACMG recommendations for reporting of incidental findings in clinical exome and genome sequencing. Genet Med 2013;15(7):565.

[33] Regier DA, Peacock SJ, Pataky R, Van Der Hoek K, Jarvik GP, Hoch J, et al. Societal preferences for the return of incidental findings from clinical genomic sequencing: a discrete-choice experiment. Can Med Assoc J 2015;187(6):E190–7.

[34] Tsiplova K, Zur RM, Marshall CR, Stavropoulos DJ, Pereira SL, Merico D, et al. A microcosting and cost-consequence analysis of clinical genomic testing strategies in autism spectrum disorder. Genet Med 2017;19 (11):1268–75.

[35] Stark Z, Schofield D, Alam K, Wilson W, Mupfeki N, Macciocca I, et al. Prospective comparison of the cost-effectiveness of clinical whole-exome sequencing with that of usual care overwhelmingly supports early use and reimbursement. Gen Med 2017;19(8):867–74.

[36] Schofield D, Rynehart L, Shresthra R, White SM, Stark Z. Long-term economic impacts of exome sequencing for suspected monogenic disorders: diagnosis, management, and reproductive outcomes. Genet Med 2019;21(11):2586–93.

[37] Yeung A, Tan NB, Tan TY, Stark Z, Brown N, Hunter MF, et al. A cost-effectiveness analysis of genomic sequencing in a prospective versus historical cohort of complex pediatric patients. Genet Med 2020;22.

[38] Regier DA, Weymann D, Buchanan J, Marshall DA, Wordsworth S. Valuation of health and nonhealth outcomes from next-generation sequencing: approaches, challenges, and solutions. Value Health 2018;21 (9):1043–7.

[39] Sagoo GS, Mohammed S, Barton G, Norbury G, Ahn JW, Ogilvie CM, et al. Cost effectiveness of using array-CGH for diagnosing learning disability. Appl Health Econ Health Policy 2015;13(4):421–32.

[40] Schofield D, Alam K, Douglas L, Shrestha R, MacArthur DG, Davis M, et al. Cost-effectiveness of massively parallel sequencing for diagnosis of paediatric muscle diseases. NPJ Genom Med 2017;2.

[41] Hayeems RZ, Bhawra J, Tsiplova K, Meyn MS, Monfared N, Bowdin S, et al. Care and cost consequences of pediatric whole genome sequencing compared to chromosome microarray. Eur J Human Genet 2017;25 (12):1303–12.

[42] CADTH. Guidelines for the economic evaluation of health technologies: Canada. Ottawa; 2017.

[43] Drummond M. Clinical guidelines: a NICE way to introduce cost-effectiveness considerations? Value Health 2016;19(5):525–30.

[44] Morisse F, Vandemaele E, Claes C, Claes L, Vandevelde S. Quality of life in persons with intellectual disabilities and mental health problems: an explorative study. TheScientificWorldJOURNAL 2013;2013, 491918.

[45] Dirita PA, Parmenter TR, Stancliffe RJ. Utility, economic rationalism and the circumscription of agency. J Intellect Disabil Res 2008;52(7):618–25.

[46] Rogers JM, Hook CC, Havyer RD. Medicine's valuing of "normal" cognitive ability. AMA J Ethics 2015;17 (8):717–26.

[47] Vissers L, van Nimwegen KJM, Schieving JH, Kamsteeg EJ, Kleefstra T, Yntema HG, et al. A clinical utility study of exome sequencing versus conventional genetic testing in pediatric neurology. Genet Med 2017;19 (9):1055–63.

[48] Soden SE, Saunders CJ, Willig LK, Farrow EG, Smith LD, Petrikin JE, et al. Effectiveness of exome and genome sequencing guided by acuity of illness for diagnosis of neurodevelopmental disorders. Sci Transl Med 2014;6(265), 265ra168.

[49] Azimi M, Schmaus K, Greger V, Neitzel D, Rochelle R, Dinh T. Carrier screening by next-generation sequencing: health benefits and cost effectiveness. Mol Genet Genom Med 2016;4(3):292–302.

[50] Grosse SD, McBride CM, Evans JP, Khoury MJ. Personal utility and genomic information: look before you leap. Genet Med 2009;11(8):575–6.
[51] Kohler JN, Turbitt E, Lewis KL, Wilfond BS, Jamal L, Peay HL, et al. Defining personal utility in genomics: a Delphi study. Clin Genet 2017;92(3):290–7.
[52] Najafzadeh M, Johnston KM, Peacock SJ, Connors JM, Marra MA, Lynd LD, et al. Genomic testing to determine drug response: measuring preferences of the public and patients using Discrete Choice Experiment (DCE). BMC Health Serv Res 2013;13:454.
[53] Versteegh MM, Brouwer WBF. Patient and general public preferences for health states: a call to reconsider current guidelines. Soc Sci Med 2016;165:66–74.
[54] Sanders GD, Neumann PJ, Basu A, Brock DW, Feeny D, Krahn M, et al. Recommendations for conduct, methodological practices, and reporting of cost-effectiveness analyses: second panel on cost-effectiveness in health and medicine. JAMA 2016;316(10):1093–103.
[55] Regier DA, Friedman JM, Makela N, Ryan M, Marra CA. Valuing the benefit of diagnostic testing for genetic causes of idiopathic developmental disability: willingness to pay from families of affected children. Clin Genet 2009;75(6):514–21.
[56] Petrou S. Methodological issues raised by preference-based approaches to measuring the health status of children. Health Econ 2003;12(8):697–702.
[57] Kwon J, Kim SW, Ungar WJ, Tsiplova K, Madan J, Petrou S. A systematic review and meta-analysis of childhood health utilities. Med Decision Making 2018;38(3):277–305.
[58] Kwon J, Kim SW, Ungar WJ, Tsiplova K, Madan J, Petrou S. Patterns, trends and methodological associations in the measurement and valuation of childhood health utilities. Qual Life Res Int J Qual Life Asp Treat Care Rehab 2019;28.
[59] Wordsworth S, Buchanan J, Regan R, Davison V, Smith K, Dyer S, et al. Diagnosing idiopathic learning disability: a cost-effectiveness analysis of microarray technology in the National Health Service of the United Kingdom. Genom Med 2007;1(1-2):35–45.
[60] Frank M, Prenzler A, Eils R, Graf von der Schulenburg JM. Genome sequencing: a systematic review of health economic evidence. Heal Econ Rev 2013;3(1):29.
[61] Regier DA, Friedman JM, Marra CA. Value for money? Array genomic hybridization for diagnostic testing for genetic causes of intellectual disability. Am J Hum Genet 2010;86(5):765–72.
[62] Regier DA, Friedman JM, Ryan M, Marra CM. Valuing the benefit of diagnostic testing of idiopathic intellectual disability: willingness to pay from families of affected children. Clinical Genetics 2009;75(6):514–21.
[63] Perez MV, Kumarasamy NA, Owens DK, Wang PJ, Hlatky MA. Cost-effectiveness of genetic testing in family members of patients with long-QT syndrome. Circul Cardiovas Quality Outcomes 2011;4(1):76–84.
[64] Tan TY, Dillon OJ, Stark Z, Schofield D, Alam K, Shrestha R, et al. Diagnostic impact and cost-effectiveness of whole-exome sequencing for ambulant children with suspected monogenic conditions. JAMA Pediatr 2017;171(9):855–62.
[65] Cordoba M, Rodriguez-Quiroga SA, Vega PA, Salinas V, Perez-Maturo J, Amartino H, et al. Whole exome sequencing in neurogenetic odysseys: an effective, cost- and time-saving diagnostic approach. PLoS One 2018;13(2), e0191228.

CHAPTER

5 Economic analysis of pharmacogenetics testing for human leukocyte antigen-based adverse drug reactions

Rika Yuliwulandari[a,b], Usa Chaikledkaew[c], Kinasih Prayuni[b], Hilyatuz Zahroh[b], Surakameth Mahasirimongkol[d], Saowalak Turongkaravee[c], Jiraphun Jittikoon[c], Sukanya Wattanapokayakit[d], and George P. Patrinos[e,f,g]

[a]*Department of Pharmacology, Faculty of Medicine, YARSI University, Jakarta, Indonesia,* [b]*Genetics Research Centre, YARSI University, Jakarta, Indonesia,* [c]*Faculty of Pharmacy, Mahidol University, Salaya, Thailand,* [d]*Ministry of Public Health, Mueang Nonthaburi, Thailand,* [e]*Department of Pharmacy, School of Health Sciences, University of Patras, Patras, Greece,* [f]*Department of Genetics and Genomics, College of Medicine and Health Sciences, United Arab Emirates University, Al-Ain, United Arab Emirates,* [g]*Zayed Center of Health Sciences, United Arab Emirates University, Al-Ain, United Arab Emirates*

5.1 Introduction

Adverse drug reactions (ADRs) occurs in relation with therapeutic dose of the drug and are a common treatment effect [1]. While some reactions are mild, more severe reactions can place significant burden on patients and healthcare systems and have been documented as the fifth leading cause of death among all diseases [1–3]. ADRs can be grouped into two categories: type A and type B. Type A reactions are predictable, common, and related to the pharmacological actions of the drug. Conversely, type B reactions are unpredictable, uncommon, and usually not related to the pharmacological actions of the drug [3,4]. Approximately 80% of ADRs are classified into the first category, including drug-induced toxicity, side effects, and drug interactions. Type B reactions involve 6% to 10% of all ADRs and include drug intolerance, idiosyncratic reactions, and hypersensitivity drug reactions (HDRs), mediated by immunological mechanisms [4].

Ethnicity is one of the key factors that influences ADRs. A systematic review by McDowell et al. suggested ethnic variations in susceptibility to ADRs due to cardiovascular medications [5]. Another example of inter-ethnic variations in susceptibility to ADRs is severe cutaneous adverse drug reactions (SCARs) that occur in some commonly used antiepileptic drugs (AEDs), antibiotics, antiviral drugs, and nonsteroidal anti-inflammatory drugs (NSAIDs) [6]. These reactions are immunologically mediated and associated with the presence of specific human leukocyte antigen (HLA) antigens. HLA genes are highly polymorphic and their frequencies in different populations vary remarkably. For example,

Economic Evaluation in Genomic and Precision Medicine. https://doi.org/10.1016/B978-0-12-813382-8.00003-3

the most common reports were found for abacavir hypersensitivity reaction (HSR) and *HLA-B*57:01*, SJS/TEN associated with carbamazepine and *HLA-B*15:02*, and allopurinol-induced Stevens-Johnson syndrome/toxic epidermal necrolysis (SJS/TEN) in patients with *HLA-B*58:01* [7,8]. Different ethnic populations may have different risks of SCARs due to genetic reasons [9]. Genetic susceptibility is a key feature of the most serious SCARs, thus there is interest in developing genetic tests to identify all those at risk of adverse events prior to prescription [10]. Some genetic associations have proven to be clinically effective and cost-effective in reducing the burden associated with serious ADRs, for example, pharmacogenetic testing of *HLA-B*57:01* before commencing abacavir for human immunodeficiency virus (HIV) in the United Kingdom [11], France [12], and Australia [13], and of *HLA-B*58:01* in Thailand [14].

ADRs also cause significant economic impacts. A landmark study of almost 20,000 patients admitted to hospital in the United Kingdom found that ADRs lead to an average of eight additional days of hospital stay and are associated with costs of approximately €706 million per year, including ADRs judged to be potentially preventable [15]. A more recent systematic review of 29 observational studies found that the incremental total cost for each patient with an ADE ranged from roughly € 702 to € 7318 [16]. ADRs lead to 6% to 7% of hospital admissions, increase length of hospital stay by two days at an increased cost of ~$2500 per patient, and increase healthcare costs. ADRs also cause drug withdrawal, which has financial consequences for pharmaceutical industries [17]. With the increasing evidence of genetic association with ADRs, healthcare practitioners and the pharmaceutical industry are actively working towards finding genetic biomarkers to alleviate patient risks to ADRs.

Studies have demonstrated that pharmacogenetic testing has the potential to avoid ADRs. With the cost reduction of genotyping in recent years, the application ADR-related genetic testing becomes an attractive concept for routine diagnostics. However, there are still some concerns about the running cost of HLA screening assays and the cost of alternative drugs if a patient is considered likely to experience an ADR if the initial drug is given. Furthermore, many healthcare stakeholders require evidence of the clinical efficacy, effectiveness, and cost-effectiveness of new healthcare technologies to provide recommendations for reimbursement, which for some countries includes pharmacogenetic tests and not just drugs [18]. On the patient side, ADR-related pharmacogenetic tests, unlike drugs, are taken once in a lifetime or prior to a treatment, hence lowering the patient's medical bills. To successfully adopt HLA genotyping into routine clinical care, health economic evidence is required to demonstrate that HLA genotyping is cost-effective compared to not using HLA genotyping. As such, this chapter provides an overview of HLA-related carbamazepine-, allopurinol-, and abacavir-induced ADRs as the most well-studied examples of clinical application of pharmacogenomics, highlighting economic evaluation information of its adaptation in clinical practice, the recommendations of regulatory authorities, and its implementation in clinical settings.

5.2 Economic evaluation of adverse drug reaction-related human leukocyte antigen genotyping, its guidelines, and current implementations

5.2.1 Abacavir-induced hypersensitivity reaction

Abacavir is a nucleoside reverse-transcriptase inhibitor that is part of highly active antiretroviral therapy (HAART) against HIV that has shown efficacy and has a favorable long-term toxicity profile [19]. The drug is available for once-daily use both as a single agent or in combination with other antiretroviral agents [20]. However, significant adverse effects of abacavir that mandate a high degree of clinical vigilance and often limit its use are immunologically mediated HSRs. These reactions can affect 5% to 8% of patients during the first 6 weeks of treatment and are characterized by fever, rash, gastrointestinal (GI) symptoms, generalized malaise, and acute respiratory problems [21]. The combination of symptoms seen in this reaction to abacavir is often dose independent. Continuing can worsen the reaction, and permanent discontinuation is warranted for patients developing HSRs, as rechallenge with abacavir can cause severe and life-threatening reactions [22,23].

An association between an HSR to abacavir and major histocompatibility complex class I allele *HLA-B*57:01* has been reported along with ethnicity-specific characteristics [21,22,24,25]. The first suggestion of genetics as a potential risk factor in predisposing abacavir HSRs came in 2002 from a study indicating a lower risk of abacavir HSRs in Black patients compared with other ethnic groups [26]. Following this, two independent studies identified a strong association between abacavir HSRs and *HLA-B*57:01* in Western Australia and in various races from North America [23,27]. Those with *HLA-B*57:01* are at increased risk of HSRs to abacavir compared to *HLA-B*57:01*. The result was replicated in UK patients [28]. The *HLA-B*57:01* is most common in Europeans (~6.8%) and US Caucasians (6%–8%) when compared with Africans (1%), South Americans (2.6%), Middle Easterners (2.5%), and Mexicans (2.2%) [23,27,29,30]. In Asians, its prevalence ranges from 0% to 0.5% among Japanese, Korean, and Taiwanese individuals to 4% and 3.4% among Thai and Cambodian individuals [30] to [30–32]. However, its association to HSRs is not restricted to race, as we will show later in our discussion of several studies that identified clinical evidence supporting the widespread utilization of *HLA-B*57:01* screening regardless of race.

Following evidence of the genetic association of *HLA-B*57*:01 with ABC HSRs, subsequent work focused on translating this discovery into clinical implementation, starting with prospective screenings to assess whether pre-treatment screening could reduce ABC HSRs. A study involving a Western Australian HIV cohort indicated that prospective genetic testing involving *HLA-B*57*:01 alone for abacavir hypersensitivity highly predicted and reduced the incidence of ABC HSRs in 1.6% of the cohort [25]. Another prospective genetic screening study in the same population also reported no cases of ABC HSRs among all *HLA-B*5:701*-negative recipients [13]. Independently, a study in ethnically mixed French population also showed that *HLA-B*57:01* testing decreased hypersensitivity from 12% to 0% as well as decreased the rate of unwarranted interruptions of abacavir treatment from 10.2% to 0.73% [12].

The first large prospective, randomized, double-blind, multicenter trial, PREDICT-1, involving around 2000 abacavir-naïve HIV patients of primarily European ancestry, reported 100% negative predictive value of *HLA-B*57:01* for abacavir HSRs and confirmed the clinical utility of *HLA-B*57:01* pre-treatment testing to avoid ABC HSRs [20]. Supporting the PREDICT-1 result, a US-based retrospective case-control study of Black and White patients called SHAPE revealed 100% negative predictive value of the *HLA-B*57:01* allele for immunologically confirmed ABC HSRs [21]. The study also showed *HLA-B*57:01* equal sensitivity as a marker for ABC HSRs in both White and Black patients in the United States despite its lower frequency and the lower incidence of ABC HSRs in US Black patients [21]. Taken together, these two studies highlight the significant association of *HLA-B*57:01* with increased risk of abacavir HSRs and the pre-therapy assessment for the presence *HLA-B*57:01* followed by avoidance of abacavir for *HLA-B*57:01*-positive patients regardless of race. These findings support the extended implementation of *HLA-B*57:01* testing in generalized clinical settings across race prior to abacavir medication.

5.2.1.1 Health economic evidence for HLA-B*57:01 screening for HIV

In terms of the health economic evidence for *HLA-B*57:01* screening for HIV-positive patients, we found five economic evaluations: two in the United States [33,34] and one each in Spain, [35], the United Kingdom [28], and Singapore [36]. Four of the studies found that *HLA-B*57:01* testing was cost-effective for preventing HSRs from abacavir when compared to no testing [28,33–35]. However, the study in Singapore found the opposite and suggested that *HLA-B*57:01* testing was unlikely to be cost-effective because of the mortality rate from abacavir-induced HSRs and genotyping costs [36]. In fact, abacavir (as first-line treatment) without genotyping was found to be the least costly and most cost-effective treatment for all ethnicities except for early-stage Indian HIV patients who contraindicated to another treatment, tenofovir. Table 5.1 summarizes the findings of the five economic evaluations.

The recommendations released by health agencies, such as the US Food and Drug Administration (FDA), the US Department of Health and Human Services (HHS), and the European Medicines Agency (EMA), are consistent with each other [29]. These health agencies recommend *HLA-B*5701* genetic screening before prescribing abacavir, which is clearly stated on the FDA-approved label for abacavir. Table 5.2 summarizes the recommended therapeutic use of abacavir in relation to *HLA-B* genotype from major health agencies across the world. The abacavir pre-therapy screening of *HLA-B*57:01*

Table 5.1 Economic evaluation studies on *HLA-B*57:01* testing before abacavir treatment.

Authors	Year of publication	Country	Price of genetic test reported	Price in 2019[a]	Study conclusions on pharmacogenetics strategy
Hughes et al. [28]	2004	UK	€43.40	USD 61,26	Cost-effective, depending on the balance between the extra costs of genotyping and the improved diagnostic test criteria
Schackman et al. [34]	2008	US	US$68.00	USD 82,37	Cost-effective only if abacavir-based treatment is as effective and costs less than tenofovir-based treatment
Kauf et al. [33]	2010	US	US$87.92	USD 102,91	Cost-saving compared with the initiation of tenofovir-containing regimen in the base case and in sensitivity analyses
Nieves Calatrava et al. [35]	2010	Spain	€55.00	USD 68,89	Economic saving (limited additional cost for the National Health System, which could be offset by lower incidence of HSRs)
Kapoor et al. [36]	2015	Singapore	US$277.00	USD 293,81	Not cost-effective (except for Indian patients with early-stage HIV who are contraindicated to tenofovir)

[a]*Note: Price adjusted with inflation rate and in USD currency.*

Table 5.2 Abacavir therapeutic recommendations released based on *HLA-B*57:01*.

Genotype	Implications	Health agency	Recommendations	References
*HLA-B*57:01*-positive	Posing a higher risk of developing ABC HSRs	United States Food and Drug Administration (US FDA), 2017	All patients should be screened for the *HLA-B*57:01* allele before commencing abacavir therapy or reinitiating abacavir therapy unless patients have a previously documented *HLA-B*57:01* allele assessment Abacavir is contraindicated in patients with a prior HSR to abacavir and in *HLA-B*57:01*-positive patients	[20,21,29]
		The Clinical Pharmacogenetics Implementation Consortium (CPIC), 2014	*HLA-B*57:01* screening should be performed in all abacavir-naive individuals before initiating abacavir-containing therapy. In abacavir-naive individuals who are *HLA-B*57:01*-positive, abacavir is not recommended and should be considered only under exceptional circumstances when the potential benefit, based on resistance patterns and treatment history, outweighs the risk	[19,29]
		EMA, 2016	Before initiating treatment with abacavir, screening for carriage of the *HLA-B*57:01* allele should be performed in any HIV-infected patient, irrespective of racial origin. Abacavir should not be used in patients known to carry the *HLA-B*57:01* allele	[29,37,38]
		The Dutch Pharmacogenetics Working Group (DPWG), 2017	Abacavir is contraindicated for *HLA-B*57:01*-positive patients. Advise the prescriber to prescribe an alternative according to current guidelines	[29]
*HLA-B*57:01*-negative	Low or reduced risk of abacavir hypersensitivity	CPIC, 2014	Use abacavir per standard dosing guidelines	[19,29]

Table adapted from Dean L. Abacavir therapy and HLA-B*57:01 *genotype. In: Pratt V, McLeod H, Rubinstein W, (Eds). Medical genetics summaries. Bethesda: National Center for Biotechnology Information (US), 2018 and Fricke-Galindo I, Llerena A, López-López M. An update on HLA alleles associated with adverse drug reactions. Drug Metab Pers Ther 2017;32(2):73–87.*

is now a standard of care in HIV treatment, especially in developed countries. In Asia, Thailand is leading in pharmacogenetics studies of antiretroviral drugs and their implementation [39].

To date, abacavir-related *HLA-B*57:01* screening remains the most commonly requested ADR-related pharmacogenetic test and perhaps the most successful model roadmap from pharmacogenetics discovery to clinical implementation [37,40].

5.2.2 Allopurinol-induced hypersensitivity reaction

Allopurinol, a xanthine oxidase inhibitor, is widely used for the treatment of gout and hyperuricemia [41]. It is generally well tolerated, effective, and low cost. However, its use is limited by allopurinol-related HSRs, which range from mild reactions to serious SCARs such as hypersensitivity syndrome and SJS/TEN [42]. Allopurinol is listed as one of the major causes of SCARs and SJS/TEN [43]. Despite its link to HSRs and the availability of alternative treatments, allopurinol is still routinely used to treat gout and hyperuricemia [41].

Previous research identified that allopurinol-induced HSRs are strongly associated with the presence of *HLA-B*58:01* in patients (odds ratio 580, 95% CI 34–9781) [44]. This association has been replicated in Caucasian [45], Japanese [46], Thai [47,48], and South Korean [49] patients, indicating that the same allele can predispose patients of different ethnicities to a serious immunological reaction with the same drug.

Prospective screening of *HLA-B*58:01* in a cohort of more than 2000 Taiwanese individuals showed that screening followed by withholding allopurinol in those with positive results reduced allopurinol-related ADRs. The study argued that pre-treatment screening could prevent around 330 incidences of allopurinol-related SCARs per 110,000 newly prescribed allopurinol users annually [50].

5.2.2.1 *Health economic evidence for* HLA-B*58:01 *screening before prescribing allopurinol*

In terms of the health economic evidence for *HLA-B*58:01* screening, we found five economic evaluation studies of *HLA-B*58:01* screening before prescribing allopurinol in gout patients to prevent SJS/TEN and drug rash with eosinophilia and systemic signs (DRESS). Individual studies were performed in the United Kingdom [51], Korea [52], Singapore [53], Malaysia [54], and Thailand [14]. Studies in Thai and Korean populations showed that *HLA-B*58:01* genotyping is a more cost-effective option than substituting allopurinol with a new xanthine oxidase inhibitor, febuxostat. [14,52]. However, the United Kingdom study suggested that *HLA-B*58:01* genotyping is not cost-effective, as the prices of the testing and febuxostat are too high [50,51]. The studies in Malaysia and Singapore also concluded that *HLA-B*58:01* was unlikely to be cost-effective compared to current practices because of the low incidence of allopurinol-induce SJS/TEN in both countries, low efficacy of the alternative drug (probenecid), and low positive predictive value for *HLA-B*58:01* testing (1.52%) [53,54]. The analysis model used in the study could be the reason for the discrepancy (e.g., the incorporation of the alternative drug efficacy in the Malaysian study) (Table 5.3) [54].

Currently there are no therapeutic recommendations for allopurinol-related HLA genotyping from the FDA and EMA [8]. Instead, recommendations were released by CPIC and the American College of Rheumatology (ACR), as listed in Table 5.2.

5.2.3 Carbamazepine-induced hypersensitivity reaction

Carbamazepine (CBZ) is widely used as a first-line treatment for epilepsy, trigeminal neuralgia, and bipolar disorder. The three common cutaneous ADRs ranging from the mildest to the most severe are (1) maculopapular exanthema (MPE), which typically does not require any specific treatment apart

Table 5.3 Economic evaluation studies on *HLA-B*58:01* testing before allopurinol treatment.

Authors	Year of publication	Country	Price of genetic test quoted in study	Price in 2019[a]	Conclusion reached by studies for pharmacogenetics strategy
Saokaew et al. [14]	2014	Thailand	1000.00	USD 33,14	Cost-effective
Park et al. [52]	2015	Korea	US$31.30	USD 33,20	Cost-effective
Dong et al. [53]	2015	Singapore	US$270.00	USD 286,38	Not cost-effective at the time of study
Plumpton et al.	2017	UK	£55.50	USD 73,95	Not cost-effective at the time of study
Chong et al. [54]	2018	Malaysia	MYR287.17	USD 69,92	Not cost-effective

[a]*Note: Price adjusted with inflation rate and in USD currency.*

from drug withdrawal, (2) hypersensitivity syndrome where the skin eruption is accompanied by systemic manifestations and extra-cutaneous involvement, most commonly the liver, and (3) SJS/TEN, which are associated with 10%–30% mortality (Table 5.4) [9].

Contrary to abacavir-induced HSRs, CBZ-induced SJS/TEN has shown a strong (OR>1000) association with *HLA-B*15:02* in the Han Chinese population, but not in Caucasian patients [45,57–59]. The association has also been replicated in several other Asian populations including Thai [60–62], Malay [63], and Indian [64], but not in Japanese [65]. The prevalence of *HLA-B*15:02* is high in Asian ethnics including Chinese, Indonesian, Malaysian, Taiwanese, Thai, and Vietnamese, but rare (<1%) in Japanese, Korean, African American, European, and Hispanic populations [66].

Besides *HLA-B*15:02*, another important association with CBZ-induced ADRs was found from the result of a genome-wide association study (GWAS) for *HLA-A*31:01*, especially in Caucasian, Japanese, and Mexican mestizos populations [67]. The allele is mainly predictive for MPE, HSS, and DRESS [37].

In terms of the cost-effectiveness of *HLA-B*15:02* genotyping, we found four economic evaluation studies carried out in Thailand (two studies) [68,69], Singapore (one study) [70], and Indonesia (one study) [70]. One of the studies in Thailand, conducted in patients with either newly diagnosed epilepsy or neuropathic pain, showed that *HLA-B*15:02* screening for CBZ-induced severe ADRs compared to CBZ therapy without screening was cost-effective in CBZ-treated patients with neuropathic pain, but not in patients with epilepsy [68]. However, the study in Singapore demonstrated that *HLA-B*15:02* genotyping for epileptic patients would be cost-effective for Singaporean Chinese and Malays, but not Indians [70]. Moreover, the other study conducted in patients receiving CBZ demonstrated that *HLA-B*15:02* screening in Thai CBZ users was cost-effective since the treatment cost of SJS/TEN was more expensive than that of *HLA-B*15:02* screening [69]. The prevalence of CBZ-induced SJS/TEN in the Thai population and the positive predictive value (PPV) were major factors affecting the cost-effectiveness results. PPV is commonly used to indicate the probability of drug-induced ADRs occurring in patients who tested positive for the risk allele. Negative predictive value (NPV) was used to indicate the probability of drug-induced ADRs occurring in patients who tested negative for the risk allele.

Table 5.4 Allopurinol therapeutic recommendations based on HLA genotype.

Genotype	Implications	Publisher	Recommendations	References
*HLA-B*58:01* positive	High risk of SCARs	CPIC, 2015	Given the high specificity for allopurinol-induced SCARs, allopurinol should not be prescribed to patients who have tested positive for *HLA-B*58:01*. Alternative medication should be considered for these patients to avoid the risk of developing SCARs. For patients who have tested negative, allopurinol may be prescribed as usual However, testing negative for *HLA-B*58:01* does not eliminate the possibility of developing SCARs, especially in the European population	[43,55,56]
*HLA-B*5801* positive	Very high hazard ratio for severe HSRs due to allopurinol	American College of Rheumatology (ACR), 2012	Prior to initiation of allopurinol, rapid, PCR-based *HLA-B*58:01* screening should be considered as a risk management component in sub-populations where both *HLA-B*58:01* allele frequency is elevated and the *HLA-B*58:01*-positive subjects have a very high hazard ratio ("high risk") for severe allopurinol HSRs (e.g., Koreans with stage 3 or worse chronic kidney disease (CKD), and all those of Han Chinese and Thai descent), including those with chronic renal insufficiency	[41,43]

*Table adapted from Khanna D, Fitzgerald JD, Khanna PP, Bae S, Singh MK, Neogi T, et al. 2012 American College of Rheumatology guidelines for management of gout. Part 1: systematic nonpharmacologic and pharmacologic therapeutic approaches to hyperuricemia. Arthritis Care Res. 2012 [cited 2018 Nov 28];64(10):1431–46. Available from: http://www.ncbi.nlm.nih.gov/pubmed/23024028, Dean L. Allopurinol therapy and HLA-B*58:01 genotype. In: Pratt VM, Scott SA, Pirmohamed M, editors. Medical genetics summaries. Bethesda: National Center for Biotechnology Information (US); 2016. Available from: https://www.ncbi.nlm.nih.gov/books/NBK127547/, Saito Y, Stamp LK, Caudle KE, Hershfield MS, McDonagh EM, Callaghan JT, et al. Clinical Pharmacogenetics Implementation Consortium (CPIC) guidelines for human leukocyte antigen B (HLA-B) genotype and allopurinol dosing: 2015 update. Clin Pharmacol Ther 2016 [cited 2018 Nov 27];99(1):36–7. Available from: http://www.ncbi.nlm.nih.gov/pubmed/26094938, Hershfield MS, Callaghan JT, Tassaneeyakul W, Mushiroda T, Thorn CF, Klein TE, et al. Clinical Pharmacogenetics Implementation Consortium guidelines for human leukocyte antigen-B genotype and allopurinol dosing. Clin Pharmacol Ther. 2013 [cited 2018 Nov 27];93(2):153–8. Available from: http://www.ncbi.nlm.nih.gov/pubmed/23232549.*

Another cost-effectiveness analysis from Indonesia using a generic model [71] demonstrated that neither *HLA-B*15:02* testing before CBZ treatment nor substitution with VPA is cost-effective in Indonesia (Table 5.5) [73]. However, considering the improvement in clinical outcomes from the implementation of *HLA-B*15:02* genetic screening in other Asian countries, this test could be potentially advantageous, especially in reducing the economic impact of CBZ-related ADRs.

Table 5.5 Economic evaluations of *HLA-B*15:02* and *HLA-A*31:01* testing before carbamazepine treatment.

Authors	Year of publication	Country	Price of genetic test quoted in study	Price in 2019[a]	Conclusion reached by studies for pharmacogenetics strategy
Rattanavipapong et al. [68]	2013	Thailand	1000	$33.87	Undetermined
Tiamkao et al. [69]	2013	Thailand	3000	$101.60	Cost-effective
Dong et al [70]	2012	Singapore	USD 270	$286.38	Cost-effective for Singaporean Chinese and Malays, but not Indians
Plumpton et al. [72]	2015	UK	£54.26	$76.20	Cost-effective
Yuliwulandari et al [73]	2021	Indonesia	IDR 1,000,000	$83.45	Not cost-effective

[a]*Note: Price adjusted with inflation rate and in USD currency.*

Following these findings, the drug label for CBZ was changed by several regulatory agencies, including the EMA and FDA. The FDA and CPIC recommend genetic screening for at-risk populations before starting CBZ therapy [58,60,74]. In 2008, the FDA informed healthcare professionals about the potential risk of SJS/TEN-related phenytoin and fosphenytoin. Therefore, healthcare providers were expected to consider avoiding phenytoin and fosphenytoin as alternatives for carbamazepine in *HLA-B*15:02*-positive patients. However, a study in Thailand did not recognize this HLA association with phenytoin [75]. In Asia, the Ministry of Public Health Thailand and the Health Sciences Authority Singapore took active roles in pharmacogenetics research, which eventually culminated in pharmacogenetic testing recommendations. CBZ-related *HLA-B*15:02* was the first gene to be widely implemented in pre-therapy genetic screening in Asia. This genetic screening prior to CBZ prescription has now became standard practice in Thailand [61], Singapore [76], Hong Kong [77], and several parts of China [39,78]. In Taiwan and Thailand, genotyping costs are covered by national health insurance, whereas in Singapore only certain patients are subsidized (Table 5.6) [8,39].

Table 5.6 Recommendations for carbamazepine therapy based on HLA genotype.

Genotype[a]	Implication	Publisher	Recommendations
*HLA-B*15:02*	High risk of SJS/TEN	US FDA, 2018	Prior to initiating CBZ therapy, testing for *HLA-B*15:02* should be performed in patients with ancestry in populations in which *HLA-B*15:02* may be present CBZ should not be used in patients positive for *HLA-B*15:02* unless the benefits clearly outweigh the risks

Continued

Table 5.6 Recommendations for carbamazepine therapy based on HLA genotype—cont'd

Genotype	Implication	Publisher	Recommendations
*HLA-A*31:01*	Hypersensitivity reactions		The risks and benefits of CBZ therapy should be weighed before considering CBZ in patients known to be positive for HLA-A*31:01
*HLA-B*15:02* positive		DPWG, 2017	Choose an alternative if possible
*HLA-A*31:01* positive			Carefully weigh the risk of DRESS and SJS/TEN against the benefits If an alternative is an option, choose an alternative
*HLA-B*15:11* positive			Carefully weigh the risk of SJS/TEN against the benefits If an alternative is an option, choose an alternative
*HLA-B*15:02* negative and *HLA-A*31:01* negative	Normal risk of carbamazepine-induced SJS/TEN, DRESS, and MPE	CPIC, 2016	Use CBZ per standard dosing guidelines (strongly recommended)[b]
*HLA-B*15:02* negative and *HLA-A*31:01* positive	Greater risk of carbamazepine-induced SJS/TEN, DRESS, and MPE		If patient is CBZ-naïve and alternative agents are available, do not use carbamazepine (strongly recommended) No recommendation can be made with respect to choosing another aromatic anticonvulsant as an alternative agent If patient is CBZ-naïve and alternative agents are not available, consider the use of CBZ with increased frequency of clinical monitoring. Discontinue therapy at first evidence of a cutaneous adverse reaction (optional) If the patient has previously used CBZ consistently for longer than three months without incidence of cutaneous adverse reactions, cautiously consider use of CBZ (optional) Previous tolerance of CBZ is not indicative of tolerance to other aromatic anticonvulsants
*HLA-B*15:02* positive[c] and any *HLA-A*31:01* genotype (or	Greater risk of carbamazepine-induced SJS/TEN		If patient is CBZ-naïve, do not use CBZ (strongly recommended) Other aromatic anticonvulsants have weaker evidence linking SJS/TEN with the *HLA-B*15:02* allele; however, caution

Table 5.6 Recommendations for carbamazepine therapy based on HLA genotype—cont'd

Genotype	Implication	Publisher	Recommendations
*HLA-A*31:01* genotype unknown)			should still be used in choosing an alternative agent If the patient has previously used CBZ consistently for longer than three months without incidence of cutaneous adverse reactions, cautiously consider use of CBZ in the future (optional) Previous tolerance of CBZ is not indicative of tolerance to other aromatic anticonvulsants
*HLA-B*15:02*		Canadian Pharmacogenomics Network for Drug Safety (CPNDS), 2014	Genetic testing for *HLA-B*15:02* is recommended for all CBZ-naive patients before initiation of CBZ therapy (Level A—strong in patients originating from populations where *HLA-B*15:02* is common, its frequency unknown, or whose origin is unknown; Level C—optional in patients originating from populations where *HLA-B*15:02* is rare)
*HLA-A*31:01*			Genetic testing for *HLA-A*31:01* is recommended for all CBZ-naive patients before initiation of CBZ therapy (Level B—moderate in all patients)

[a]*If only* HLA-B*15:02 *was tested, assume* HLA-A**31:01 is negative and vice versa.*
[b]HLA-B*15:02 *has a 100% negative predictive value for CBZ-induced SJS/TEN, and its use is currently recommended to guide the use of carbamazepine and oxcarbazepine only. Because there is a much weaker association and less than 100% negative predictive value of* HLA-B*15:02 *for SJS/TEN associated with other aromatic anticonvulsants, using these drugs instead of carbamazepine or oxcarbazepine in the setting of a negative* HLA-B*15:02 *test in Southeast Asians will not result in prevention of anticonvulsant-associated SJS/TEN.*
[c]*In addition to* HLA-B*15:02, *the risk for CBZ-induced SJS/TEN has been reported in association with the most common B75 serotype alleles in Southeast Asia,* HLA-B**15:08*, HLA-B*15:11, *and* HLA-B*15:21. *Although not described, the possibility of CBZ-induced SJS/TEN in association with less frequently carried B75 serotype alleles, such as* HLA-B*15:30 *and* HLA-B*15:31, *should also be considered.*
*This table is adapted from Phillips EJ, Sukasem C, Whirl-Carrillo M, Müller DJ, Dunnenberger HM, Chantratita W, et al. Clinical Pharmacogenetics Implementation Consortium (CPIC) guideline for HLA genotype and use of carbamazepine and oxcarbazepine: 2017 update. Clin Pharmacol Ther 2018;103(4):574–81, Dean L. Carbamazepine therapy and HLA genotype. In: Pratt VM, Scott SA, Pirmohamed M editors. Medical genetics summaries. Bethesda: National Center for Biotechnology Information (US); 2018. Available from: https://www.ncbi.nlm.nih.gov/books/NBK321445/, Amstutz U, Shear NH, Rieder MJ, Hwang S, Fung V, Nakamura H, et al. Recommendations for HLA-B*15:02 and HLA-A*31:01 genetic testing to reduce the risk of carbamazepine-induced hypersensitivity reactions. Epilepsia. 2014 [cited 2018 Nov 27];55(4):496–506. Available from: http://www.ncbi.nlm.nih.gov/pubmed/24597466.*

5.3 Conclusions

Most of the economic evaluations reported in this chapter have suggested that pre-emptive genetic screening for HLA-based ADRs is likely to be cost-effective when compared against non-screening strategies, except in the case of allopurinol-induced ADRs, which perhaps explains why the FDA and EMA have not released pharmacogenetics-guided treatment recommendations related to allopurinol. Meanwhile, patients have been saved from severe adverse events related to abacavir and carbamazepine with the routine practice of pre-treatment HLA screening in most developed countries and in some parts of Asia, including Korea, Japan, China, Taiwan, Hong Kong, Thailand, and Singapore. More research is required to explore the wider use of pharmacogenetic tests in other disease areas. However, the clinical guidelines for the implementation of ADR-related pharmacogenetic testing have been released by the CPIC.

To implement pharmacogenetic testing as a part of precision medicine practice, it is necessary to integrate genetic data into electronic health records to allow doctors to create personalized treatment plans.

References

[1] Rive CM, Bourke J, Phillips EJ. Testing for drug hypersensitivity syndromes. Clin Biochem Rev 2013; 34(1):15–38.

[2] Verma R, Vasudevan B, Pragasam V. Severe cutaneous adverse drug reactions. Med J Armed Forces India 2013;69(4):375–83.

[3] Pavlos R, Mallal S, Phillips E. HLA and pharmacogenetics of drug hypersensitivity. Pharmacogenomics 2012;13(11):1285–306.

[4] Doña I, Barrionuevo E, Blanca-Lopez N, Torres MJ, Fernandez TD, Mayorga C, et al. Trends in hypersensitivity drug reactions: more drugs, more response patterns, more heterogeneity. J Investig Allergol Clin Immunol 2014;24(3):143–53.

[5] McDowell SE, Coleman JJ, Ferner RE. Systematic review and meta-analysis of ethnic differences in risks of adverse reactions to drugs used in cardiovascular medicine. Br Med J 2006;332:1177–80. [cited 2021 Apr 7]. Available from: /pmc/articles/PMC1463974/.

[6] Sukasem C, Puangpetch A, Medhasi S, Tassa. Pharmacogenomics of drug-induced hypersensitivity reactions: challenges, opportunities and clinical implementation. Asian Pac J Allergy Immunol 2014;32: 111–23.

[7] Roujeau JC, Stern RS. Severe adverse cutaneous reaction to drugs. N Engl J Med 1994;331(19):1272–85.

[8] Plumpton CO, Roberts D, Pirmohamed M, Hughes D, a. A systematic review of economic evaluations of pharmacogenetic testing for prevention of adverse drug reactions. PharmacoEconomics 2016;34(8):771–93.

[9] Pirmohamed M. Genetics and the potential for predictive tests in adverse drug reactions. Advers Cutan Drug Eruptions 2012;97:18–31.

[10] Karlin E, Phillips E. Genotyping for severe drug hypersensitivity. Curr Allergy Asthma Rep 2014;14(3): 1–20.

[11] Waters LJ, Mandalia S, Gazzard B, Nelson M. Prospective HLA-B*5701 screening and abacavir hypersensitivity: a single centre experience. AIDS 2007;21:2533–49.

[12] Zucman D, De TP, Majerholc C, Stegman S, Caillat-Zucman S. Prospective screening for human leukocyte antigen-B*5701 avoids abacavir hypersensitivity reaction in the ethnically mixed French HIV population. J Acquir Immune Defic Syndr 2007;45(1):1–3.

[13] Rauch A, Nolan D, Martin A, McKinnon E, Almeida C, Mallal S. Prospective genetic screening decreases the incidence of abacavir hypersensitivity reactions in the Western Australian HIV Cohort Study. Clin Infect Dis 2006;43(1):99–102. Available from: https://academic.oup.com/cid/article-lookup/doi/10.1086/504874.

[14] Saokaew S, Tassaneeyakul W, Maenthaisong R, Chaiyakunapruk N. Cost-effectiveness analysis of HLA-B*5801 testing in preventing allopurinol-induced SJS/TEN in Thai population. PLoS One 2014; 9(4):1–9.

[15] Pirmohamed M, James S, Meakin S, Green C, Scott AK, Walley TJ, et al. Adverse drug reactions as cause of admission to hospital: prospective analysis of 18,820 patients. Br Med J 2004;329(7456):15–9. [cited 2021 Apr 8]. Available from: http://www.bmj.com/.

[16] Marques FB, Penedones A, Mendes D, Alves C. A systematic review of observational studies evaluating costs of adverse drug reactions. In: Clinicoeconomics and outcomes research, vol. 8. Dove Medical Press Ltd; 2016. p. 413–26. [cited 2021 Apr 8]. Available from: /pmc/articles/PMC5003513/.

[17] Pirmohamed M, Park BK. Genetic susceptibility to adverse drug reactions. Trends Pharmacol Sci 2001; 22(6):298–305.

[18] De Leon J. Pharmacogenomics: the promise of personalized medicine for CNS disorders. Neuropsychopharmacology 2009;34(1):159–72.

[19] Martin M, Klein T, Dong B, Pirmohamed M, Haas D, Kroetz D. Clinical pharmacogenetics implementation consortium guidelines for HLA-B genotype and Abacavir Dosing. Clin Pharmacol Ther 2015;98(1):19–24. Available from: https://doi.org/10.1038/clpt.2011.355/nature06264.

[20] Mallal S, Phillips E, Carosi G, Molina J-M, Workman C, Tomažič J, et al. HLA-B*5701 screening for hypersensitivity to Abacavir. N Engl J Med 2008;358:568–79. Available from: http://pediatrics.aappublications.org/cgi/doi/10.1542/peds.2008-2139LL.

[21] Saag M, Balu R, Phillips E, Brachman P, Martorell C, Burman W, et al. High sensitivity of human leukocyte antigen-B*5701 as a marker for immunologically confirmed abacavir hypersensitivity in white and black patients. Clin Infect Dis 2008;46(7):1111–8. Available from: https://academic.oup.com/cid/article-lookup/doi/10.1086/529382.

[22] Hetherington S, McGuirk S, Powell G, Cutrell A, Naderer O, Spreen B, et al. Hypersensitivity reactions during therapy with the nucleoside reverse transcriptase inhibitor abacavir. Clin Ther 2001;23(10): 1603–14.

[23] Mallal S, Nolan D, Witt C, Masel G, Martin AM, Moore C, et al. Association between presence of HLA-B*5702, HLA-DR7, and HLA-DQ3 and hypersensitivity to HIV-1 reverse-transcriptase inhibitor abacavir. Lancet 2002;359:727–32.

[24] Mallal S, Nolan D, Witt C, Masel G, Martin AM, Moore C, et al. Association between presence of HLA-B*5701, HLA-DR7, and HLA-DQ3 and hypersensitivity to HIV-1 reverse-transcriptase inhibitor abacavir. Lancet 2002;359:727–32.

[25] Martin AM, Nolan D, Gaudieri S, Almeida CA, Nolan R, James I, et al. Predisposition to abacavir hypersensitivity conferred by HLA-B*5701 and a haplotypic Hsp70-Hom variant. Proc Natl Acad Sci 2004;101(12):4180–5. Available from: http://www.pnas.org/cgi/doi/10.1073/pnas.0307067101.

[26] Symonds W, Cutrell A, Edwards M, Steel H, Spreen B, Powell G, et al. Risk factor analysis of hypersensitivity reactions to abacavir. Clin Ther 2002;24(4):565–73. [cited 2018 Nov 25]. Available from: http://www.ncbi.nlm.nih.gov/pubmed/12017401.

[27] Hetherington S, Hughes AR, Mosteller M, Shortino D, Baker KL, Spreen W, et al. Genetic variations in HLA-B region and hypersensitivity reactions to abacavir. Lancet 2002;359(9312):1121–2. [cited 2018 Nov 25]. Available from: http://www.ncbi.nlm.nih.gov/pubmed/11943262.

[28] Hughes DA, Vilar FJ, Ward CC, Alfirevic A, Park BK, Pirmohamed M. Cost-effectiveness analysis of HLA B*5701 genotyping in preventing abacavir hypersensitivity. Pharmacogenetics 2004;14(6):335–42.

[29] Dean L. Abacavir therapy and HLA-B*57:01 genotype. In: Pratt V, McLeod H, Rubinstein W, editors. Medical genetics summaries. Bethesda: National Center for Biotechnology Information (US); 2018.

[30] Puthanakit T, Bunupuradah T, Kosalaraksa P, Vibol U, Hansudewechakul R, Ubolyam S, et al. Prevalence of human leukocyte antigen-B*5701 among HIV-infected children in Thailand and Cambodia: implications for abacavir use. Pediatr Infect Dis J 2013;32(3):252–3. [cited 2018 Nov 26]. Available from: http://www.ncbi.nlm.nih.gov/pubmed/22986704.

[31] Park WB, Choe PG, Song K, Lee S, Jang H, Jeon JH, et al. Should HLA-B*5701 screening be performed in every ethnic group before starting abacavir? Clin Infect Dis 2009;48(3):365–7. [cited 2018 Nov 10]. Available from: https://academic.oup.com/cid/article-lookup/doi/10.1086/595890.

[32] Sun H-Y, Hung C-C, Lin P-H, Chang S-F, Yang C-Y, Chang S-Y, et al. Incidence of abacavir hypersensitivity and its relationship with HLA-B*5701 in HIV-infected patients in Taiwan. J Antimicrob Chemother 2007;60(3):599–604. [cited 2018 Nov 26]. Available from: http://www.ncbi.nlm.nih.gov/pubmed/17631508.
[33] Kauf TL, Farkouh RA, Earnshaw SR, Watson ME, Maroudas P, Chambers MG. Economic efficiency of genetic screening to inform the use of abacavir sulfate in the treatment of HIV. PharmacoEconomics 2010; 28(11):1025–39.
[34] Schackman BR, Scott C, Walensky RP, Losina E, Freedberg KA, Sax PE. The cost-effectiveness of HLA-B*5701 genetic screening to guide initial antiretroviral therapy for HIV. AIDS 2008;22(15):2025–33.
[35] Nieves Calatrava D, Calle-Martín ÓD, Iribarren-Loyarte JA, Rivero-Román A, García-Bujalance L, Pérez-Escolano I, et al. Cost-effectiveness analysis of HLA-B*5701 typing in the prevention of hypersensitivity to abacavir in HIV patients in Spain. Enferm Infecc Microbiol Clin 2010;28(9):590–5.
[36] Kapoor R, Martinez-Vega R, Dong D, Tan SY, Leo YS, Lee CC, et al. Reducing hypersensitivity reactions with HLA-B*5701 genotyping before abacavir prescription: clinically useful but is it cost-effective in Singapore? Pharmacogenet Genomics 2015;25(2):60–72.
[37] Fricke-Galindo I, Llerena A, López-López M. An update on HLA alleles associated with adverse drug reactions. Drug Metab Pers Ther 2017;32(2):73–87.
[38] European Medicines Agency. Ziagen: EPAR summary for the public. vol. 44; 2016. London. Available from: https://www.ema.europa.eu/medicines/human/EPAR/ziagen.
[39] Ang HX, Chan SL, Sani LL, Quah CB, Brunham LR, Tan BOP, et al. Pharmacogenomics in Asia: a systematic review on current trends and novel discoveries. Pharmacogenomics 2017;18(9):891–910. [cited 2018 Nov 27]. Available from: https://www.futuremedicine.com/doi/10.2217/pgs-2017-0009.
[40] van Schaik RH, IFCC Task Force on Pharmacogenetics: Prof. Dr. Maurizio Ferrari (IT), Prof. Dr. Michael Neumaier (GER), Prof. Dr. Munir Pirmohamed (UK), Prof. Dr. Henk-Jan Guchelaar (NL) PDR van S (NL). Clinical application of pharmacogenetics: where are we now? EJIFCC 2013;24(3):105–12. [cited 2018 Nov 21]. Available from: http://www.ncbi.nlm.nih.gov/pubmed/27683445.
[41] Khanna D, Fitzgerald JD, Khanna PP, Bae S, Singh MK, Neogi T, et al. American College of Rheumatology guidelines for management of gout. Part 1: systematic nonpharmacologic and pharmacologic therapeutic approaches to hyperuricemia. Arthritis Care Res 2012;64(10):1431–46. [cited 2018 Nov 28]. Available from: http://www.ncbi.nlm.nih.gov/pubmed/23024028.
[42] Phillips EJ, Chung WH, Mockenhaupt M, Roujeau JC, Mallal S. Drug hypersensitivity: pharmacogenetics and clinical syndromes. J Allergy Clin Immunol 2011;127(3 Suppl):S60–6. Available from: https://doi.org/10.1016/j.jaci.2010.11.046.
[43] Dean L. Allopurinol therapy and HLA-B*58:01 genotype. In: Pratt VM, Scott SA, Pirmohamed M, editors. Medical genetics summaries. Bethesda: National Center for Biotechnology Information (US); 2016. Available from: https://www.ncbi.nlm.nih.gov/books/NBK127547/.
[44] Hung S-I, Chung W-H, Liou L-B, Chu C-C, Lin M, Huang H-P, et al. HLA-B*5801 allele as a genetic marker for severe cutaneous adverse reactions caused by allopurinol. Proc Natl Acad Sci 2005;102(11):4134–9.
[45] Lonjou C, Borot N, Sekula P, Ledger N, Thomas L, Halevy S, et al. A European study of HLA-B in Stevens-Johnson syndrome and toxic epidermal necrolysis related to five high-risk drugs. Pharmacogenet Genomics 2008;18(2):99–107.
[46] Dainichi T, Uchi H, Moroi Y, Furue M. Stevens-Johnson syndrome, drug-induced hypersensitivity syndrome and toxic epidermal necrolysis caused by allopurinol in patients with a common HLA allele: what causes the diversity? Dermatology 2007;215(1):86–8.
[47] Tassaneeyakul WW, Jantararoungtong T, Chen P, Lin PPY, Tiamkao S, Khunarkornsiri U, et al. Strong association between HLA-B*5801 and allopurinol-induced Stevens-Johnson syndrome and toxic epidermal necrolysis in a Thai population. Pharmacogenet Genomics 2009;19(9):704–9.
[48] Somkrua R, Eickman EE, Saokaew S, Lohitnavy M, Chaiyakunapruk N. Association of HLA-B*5801 allele and allopurinol-induced Stevens Johnson syndrome and toxic epidermal necrolysis: a systematic review and

meta-analysis. BMC Med Genet 2011;12(1):118. Available from: http://www.pubmedcentral.nih.gov/articlerender.fcgi?artid=3189112&tool=pmcentrez&rendertype=abstract.

[49] Kim E, Seol JE, Choi J, Kim N, Shin J. Allopurinol-induced severe cutaneous adverse reactions: a report of three cases with the *HLA-B*58:01* allele who underwent lymphocyte activation test. Transl Clin Pharmacol 2017;25(2):63–6.

[50] Ko TM, Tsai CY, Chen SY, Chen KS, Yu KH, Chu CS, et al. Use of HLA-B*58:01 genotyping to prevent allopurinol induced severe cutaneous adverse reactions in Taiwan: National Prospective Cohort Study. BMJ 2015;351:1–7.

[51] Plumpton CO, Alfirevic A, Pirmohamed M, Hughes DA. Cost effectiveness analysis of HLA-B*58:01 genotyping prior to initiation of allopurinol for gout. Rheumatology (United Kingdom) 2017;56(10):1729–39.

[52] Park DJ, Kang JH, Lee JW, Lee KE, Wen L, Kim TJ, et al. Cost-effectiveness analysis of HLA-B5801 genotyping in the treatment of gout patients with chronic renal insufficiency in Korea. Arthritis Care Res 2015; 67(2):280–7.

[53] Dong D, Tan-Koi WC, Teng GG, Finkelstein E, Sung C. Cost-effectiveness analysis of genotyping for HLA-B*5801 and an enhanced safety program in gout patients starting allopurinol in Singapore. Pharmacogenomics 2015;16(16):1781–93.

[54] Chong HY, Lim YH, Prawjaeng J, Tassaneeyakul W, Mohamed Z, Chaiyakunapruk N. Cost-effectiveness analysis of HLA-B*58:01 genetic testing before initiation of allopurinol therapy to prevent allopurinol-induced Stevens-Johnson syndrome/toxic epidermal necrolysis in a Malaysian population. Pharmacogenet Genomics 2018;28(2):56–67.

[55] Saito Y, Stamp LK, Caudle KE, Hershfield MS, McDonagh EM, Callaghan JT, et al. Clinical Pharmacogenetics Implementation Consortium (CPIC) guidelines for human leukocyte antigen B (HLA-B) genotype and allopurinol dosing: 2015 update. Clin Pharmacol Ther 2016;99(1):36–7. [cited 2018 Nov 27]. Available from: http://www.ncbi.nlm.nih.gov/pubmed/26094938.

[56] Hershfield MS, Callaghan JT, Tassaneeyakul W, Mushiroda T, Thorn CF, Klein TE, et al. Clinical Pharmacogenetics Implementation Consortium guidelines for human leukocyte antigen-B genotype and allopurinol dosing. Clin Pharmacol Ther 2013;93(2):153–8. [cited 2018 Nov 27]. Available from: http://www.ncbi.nlm.nih.gov/pubmed/23232549.

[57] Chung W-H, Hung S-I, Hong H-S, Hsih M-S, Yang L-C, Ho H-C, et al. Medical genetics: a marker for Stevens-Johnson syndrome. Nature 2004;428(6982):486. [cited 2015 Apr 21]. Available from: https://doi.org/10.1038/428486a.

[58] Man CBL, Kwan P, Baum L, Yu E, Lau KM, Cheng ASH, et al. Association between HLA-B*1502 allele and antiepileptic drug-induced cutaneous reactions in Han Chinese. Epilepsia 2007;48(5):1015–8.

[59] Wang Q, Zhou JQ, Zhou LM, Chen ZY, Fang ZY, Da CS, et al. Association between HLA-B*1502 allele and carbamazepine-induced severe cutaneous adverse reactions in Han people of southern China mainland. Seizure 2011;20(6):446–8. Available from: https://doi.org/10.1016/j.seizure.2011.02.003.

[60] Locharernkul C, Loplumlert J, Limotai C, Korkij W, Desudchit T, Tongkobpetch S, et al. Carbamazepine and phenytoin induced Stevens-Johnson syndrome is associated with HLA-B*1502 allele in Thai population. Epilepsia 2008;49(12):2087–91.

[61] Tassaneeyakul W, Tiamkao S, Jantararoungtong T, Chen P, Lin SY, Chen WH, et al. Association between HLA-B*1502 and carbamazepine-induced severe cutaneous adverse drug reactions in a Thai population. Epilepsia 2010;51(5):926–30.

[62] Tangamornsuksan W, Chaiyakunapruk N, Somkrua R, Lohitnavy M, Tassaneeyakul W. Relationship between the HLA-B*1502 allele and carbamazepine-induced Stevens-Johnson syndrome and toxic epidermal necrolysis: a systematic review and meta-analysis. JAMA Dermatol 2013;149(9):1025–32.

[63] Chang C, Too C, Murad S, Hussein SH. Pharmacology and therapeutics Association of HLA-B*1502 allele with carbamazepine-induced toxic epidermal necrolysis and Stevens-Johnson syndrome in the multi-ethnic Malaysian population. Int J Dermatol 2011;50:221–4.
[64] Mehta TY, Prajapati LM, Mittal B, Joshi CG, Sheth JJ, Patel DB, et al. Association of HLA-B*1502 allele and carbamazepine-induced Stevens-Johnson syndrome among Indians. Indian J Dermatol Venereol Leprol 2009;75(6):579–82.
[65] Kaniwa N, Saito Y, Aihara M, Matsunaga K, Tohkin M, Kurose K, et al. HLA-B*1511 is a risk factor for carbamazepine-induced Stevens-Johnson syndrome and toxic epidermal necrolysis in Japanese patients. Epilepsia 2010;51(12):2461–5.
[66] Lee JW, Aminkeng F, Bhavsar P, Shaw K, Carleton BC, Hayden MR, et al. The emerging era of pharmacogenomics: current successes, future potential, and challenges. Clin Genet 2014;86(1):21–8.
[67] Aihara M. Pharmacogenetics of cutaneous adverse drug reactions. J Dermatol 2011;38(3):246–54.
[68] Rattanavipapong W, Koopitakkajorn T, Praditsitthikorn N, Mahasirimongkol S, Teerawattananon Y. Economic evaluation of HLA-B*15:02 screening for carbamazepine-induced severe adverse drug reactions in Thailand. Epilepsia 2013;54(9):1628–38.
[69] Tiamkao S, Jitpimolmard J, Sawanyawisuth K, Jitpimolmard S. Cost minimization of HLA-B*1502 screening before prescribing carbamazepine in Thailand. Int J Clin Pharm 2013;35(4):608–12.
[70] Di D, Sung C, Finkelstein E. Cost-effectiveness of HLA-B*1502 genotyping in adult patients with newly diagnosed epilepsy in Singapore (Provisional abstract). Neurology 2012;79:1259–67. Available from: http://cochranelibrary-wiley.com/o/cochrane/cleed/articles/NHSEED-22013017350/frame.html.
[71] Snyder SR, Hao J, Cavallari LH, Geng Z, Elsey A, Johnson JA, et al. Generic cost-effectiveness models: a proof of concept of a tool for informed decision-making for public health precision medicine. Public Health Genom 2018;21(5–6):217–27. [cited 2021 Apr 8]. Available from: https://www.karger.com/Article/FullText/500725.
[72] Plumpton CO, Yip VLM, Alfirevic A, Marson AG, Pirmohamed M, Hughes DA. Cost-effectiveness of screening for HLA-A*31:01 prior to initiation of carbamazepine in epilepsy. Epilepsia 2015;56(4):556–63. [cited 2018 Nov 28]. Available from: http://www.ncbi.nlm.nih.gov/pubmed/26046144.
[73] Yuliwulandari R, Shin JG, Kristin E, Suyatna FD, Prahasto ID, Prayuni K, et al. Cost-effectiveness analysis of genotyping for HLA-B*15:02 in Indonesian patients with epilepsy using a generic model. Pharm J 2021;1–8. [cited 2021 Apr 8]. Available from: http://www.nature.com/articles/s41397-021-00225-9.
[74] Medicines and Healthcare Products Regulatory Agency. Carbamazepine, oxcarbazepine and eslicarbazepine: potential risk of serious skin reactions; 2012.
[75] Tassaneeyakul W, Prabmeechai N, Sukasem C, Kongpan T, Konyoung P, Chumworathayi P, et al. Associations between HLA class I and cytochrome P450 2C9 genetic polymorphisms and phenytoin-related severe cutaneous adverse reactions in a Thai population. Pharmacogenet Genomics 2016;26(5):225–34. [cited 2018 Nov 28]. Available from: http://www.ncbi.nlm.nih.gov/pubmed/26928377.
[76] Toh DSL, Tan LL, Aw DCW, Pang SM, Lim SH, Thirumoorthy T, et al. Building pharmacogenetics into a pharmacovigilance program in Singapore: using serious skin rash as a pilot study. Pharm J 2014;14(4):316–21. [cited 2018 Nov 27]. Available from: http://www.ncbi.nlm.nih.gov/pubmed/24394201.
[77] Chen Z, Liew D, Kwan P. Effects of a HLA-B*15:02 screening policy on antiepileptic drug use and severe skin reactions. Neurology 2014;83(22):2077–84. [cited 2018 Nov 27]. Available from: http://www.ncbi.nlm.nih.gov/pubmed/25355835.
[78] Su S-C, Hung S-I, Fan W-L, Dao R-L, Chung W-H. Severe cutaneous adverse reactions: the pharmacogenomics from research to clinical implementation. Int J Mol Sci 2016;17(11). [cited 2018 Nov 27]. Available from: http://www.ncbi.nlm.nih.gov/pubmed/27854302.

CHAPTER

6 Economic evaluation of personalized medicine interventions in medium- and low-income countries with poor proliferation of genomics and genetic testing

Christina Mitropoulou[a,b] and George P. Patrinos[b,c,d]

[a]*The Golden Helix Foundation, London, United Kingdom,* [b]*Department of Genetics and Genomics, College of Medicine and Health Sciences, United Arab Emirates University, Al-Ain, United Arab Emirates,* [c]*Department of Pharmacy, School of Health Sciences, University of Patras, Patras, Greece,* [d]*Zayed Center of Health Sciences, United Arab Emirates University, Al-Ain, United Arab Emirates*

6.1 Introduction

Pharmacogenomics focuses on the use of individual genomic information to inform clinical care (e.g., for diagnostic or therapeutic decision-making), and the health outcomes and policy implications of the use of genomic information in this way. To date, multiple clinical studies have concluded that it is beneficial to apply pharmacogenomics to guide patient stratification in fields such as oncology, cardiology, rare and undiagnosed diseases, and infectious diseases (Ref. [1] and chapters therein). Given this, both the United States Food and Drug Administration (FDA; www.fda.gov) and the European Medicines Agency (EMA; www.ema.europa.eu) have added pharmacogenomics labeling to more than 200 commonly prescribed drugs (outlined in Ref. [2]).

One may say that the goal of any healthcare system is to provide high-quality health services to meet the needs of a defined population on an equal basis, to allow quick access to innovation that improves value, and to do this efficiently by consuming as few resources as possible. In most developed countries, these health system activities are organized centrally. However, there are several obstacles to achieving these goals, including demographic problems, the cost of expensive health technologies, unhealthy lifestyles, medical errors, supplier-induced demand for services, and public expectations.

Given this, in many countries, governments or health insurers apply health technology assessment (economic evaluation) in health care [3]. Economic evaluation, commonly operationalized as cost-effectiveness analysis, offers a way of estimating the additional value to society of a new intervention (such as a genome-guided therapy) relative to current practice and helps to ensure that scarce healthcare resources are used as efficiently as possible [4,5].

Economic Evaluation in Genomic and Precision Medicine. https://doi.org/10.1016/B978-0-12-813382-8.00004-5

Table 6.1 List of countries in which genetic and genomic tests are in widespread use. These data are based on the best available information obtained from the literature, the Internet, and personal communications. However, it should be noted that this table is illustrative and thus may not include all relevant countries.

Continent	Countries
Americas	Brazil Canada United States
Europe	Denmark Estonia France Germany Italy The Netherlands Sweden Switzerland United Kingdom
Middle East	Israel Qatar Saudi Arabia
Asia-Pacific	Australia China Japan

In this chapter, we review economic evaluation studies conducted in medium- and low-income countries around the world, especially those with poor proliferation of genetic testing services and genomics. We use the income criterion to categorize these countries (https://data.worldbank.org), being a more straightforward classification criterion, but we also take into account the existing genomics knowledge and awareness as well as the available infrastructure for genetic analyses and the degree of integration of genomics technologies and proliferation of genetic testing services in each country (Table 6.1) (see also Phillips et al. [6]). Using high-income European countries as a model of clinical application in the first paragraph, we find that economic evaluation studies in pharmacogenomics are still in their infancy for the majority of the low-resourced and medium- and low-income European countries, highlighting the need to perform additional studies in these environments not only to enrich the evidence base for genomic medicine interventions but also to boost clinical implementation of genome-guided therapeutic modalities. We conclude by describing a generic economic model for measuring cost-effectiveness of genome-guided drug treatment interventions in medium- and low-income countries.

6.2 Implementing personalized medicine interventions beyond high-income countries: Setting the scene

In all countries around the world, significant scientific, legal, ethical, political, and economic challenges will need to be overcome for personalized medicine to ultimately improve population health. This will require innovative collaborations between various stakeholders, including healthcare providers, universities, and non-governmental and international organizations. These challenges are particularly acute in the health systems of low-income and lower middle-income countries outside Europe, in which implementation of genomic medicine application is at a very low level. Specific barriers also exist in the health systems of many upper middle-income countries in Europe. In this section, we set the scene for this chapter by describing some of the challenges in these two settings that pertain to economic evaluation of genome-guided interventions.

6.2.1 Challenges in low-income and lower middle-income countries outside Europe

A fundamental challenge in many low-income and lower middle-income countries is poor or non-existent healthcare delivery infrastructure, which limits access to preventive and curative care for broad populations. For example, most of the 1.2 billion people living in India lack access to health care; only the very wealthy can afford to visit private hospitals equipped with the latest imaging and medical devices [7]. Achieving universal health coverage in these economies is currently a high priority for the global community to ensure that such countries will grow sustainably in the long term. Some countries, such as China, have made promising progress in this direction, but many are struggling to keep up with the rising population burden [8].

A further challenge in these countries is that they often lack the data that is required to allocate healthcare resources rationally. Such decisions are commonly made in high-income countries by health technology assessment agencies, using data on the efficacy, safety costs, health outcomes, and cost-effectiveness of new healthcare interventions and their comparators, but such data is commonly lacking in low-income and lower middle-income countries. Drug efficacy and/or toxicity data are mostly generated in European or Anglo-American populations even though local data is essential to make rational healthcare resource allocation decisions [9]. A recent systematic review and meta-analysis found considerable variation among different racial/ethnic populations in the relationship between *HLA-B*1502* and carbamazepine (CBZ)-induced Stevens Johnson Syndrome (SJS) and Toxic Epidermal Necrolysis (TEN), with screening warranted in only three racial/ethnic subgroups (Han-Chinese, Thai, Malaysian) [10]. An intra-country example is provided by a cost-effectiveness analysis of three ethnic groups in Singapore with different allele frequencies of *HLA-B*1502*. Genotyping for *HLA-B*1502* and providing alternate medications to those who test positive was found to be cost-effective for Singaporean Chinese and Malays, but not for Singaporean Indians [11].

Of course, apart from health coverage and the lack of data for economic evaluation analyses, which constitute two majors challenges to implement personalized medicine in low-resource healthcare settings, there are also other obstacles and challenges, such as addressing poor genomics literacy, raising the public's genomics awareness, capacity building, and so on. At the same time, other challenges in this context in low-income and lower middle-income countries, especially outside Europe, could relate to inadequate or non-existent institutions to perform health technology assessment (HTA), such as well-structured HTA organizations, or lack of specialist expertise to conduct economic evaluations. Detailed analysis of these challenges lies outside the scope of the present chapter.

6.2.2 Challenges in upper middle-income countries in Europe

Despite advances in research and clinical implementation in high-income European countries (e.g., the United Kingdom, the Netherlands, Switzerland, Germany, and the Scandinavian countries), the clinical application of pharmacogenomics and personalized medicine is still in its infancy for many upper middle-income countries in Europe. As noted in Chapters 1 and 3, data on cost-effectiveness is required if pharmacogenomics is to be implemented in the clinic in these countries (further information on cost-effectiveness analysis is provided in Section 1.3). However, cost-effectiveness studies are mainly undertaken in high-income countries, even though one could argue that such analyses are much more important to lower-income countries where the opportunity cost of allocating scarce healthcare resources to technology that is not cost-effective is high. There is also a shortage of health economists and modelers in these countries, especially with expertise in personalized medicine.

In addition to challenges related to evidence development, it is also important in these countries to account for the variable prevalence and unique allelic spectrum of pharmacogenomic biomarkers when evaluating the cost-effectiveness of personalized medicine interventions. Indeed, it is possible for the cost-effectiveness of a pharmacogenomics-guided medical intervention to vary between two countries, even if no significant cost differences exist, due to differences in the frequency of a pharmacogenomic biomarker in the general population. Such genomic variation can also exist between geographical regions in the same country [12]. For example, the Russian population clusters into several large ethno-geographical groups (Slavs, Northern Caucasus populations, Finno-Ugric people of North European and Volga-Ural regions, the populations of South Siberia and Central Asia, the populations of Eastern Siberia and North Asia).

Several examples exist of alleles that differ in frequency in different populations. For example, the *CYP2D6*7* allele is more common in the Maltese population compared to the overall Caucasian population, and the same is true for the frequency of the *CYP2C9*3* allele in the Serbian population [12], which could have significant implications for the cost-effectiveness of treatments such as risperidone and warfarin, respectively, in these two countries. Another example is screening for the *HLA-B*1502* allele, which elevates patients' risk of developing SJS and TEN when treated with the antiepileptic drugs (AEDs) carbamazepine (CBZ) and phenytoin (PHT).

Another example using data from high-income countries involves *CYP2C9* and *VKORC1* genotyping-guided treatment with coumarin derivatives (e.g., acenocoumarol, phenprocoumon) that considered the cost of anticoagulant care management in the Netherlands. This study identified variations between countries in the setting of International Normalized Ratio (INR) monitoring and coumarin dosing, the frequency of INR monitoring, and in the prevalence of coumarin use. From the reasons described, it is clear that in countries where anticoagulant care is less well organized, there is the highest probability for pharmacogenomics to be cost-effective. The reason for that is the lack of well-organized anticoagulation clinics that increases the need of genome-guided treatment, which in turn reduces adverse drug reactions, leading to a greater potential to save healthcare costs and improve the quality of life of patients. Nevertheless, genotyping might still be a cost-effective strategy in countries where anticoagulant care is well organized, as fewer INR measurements would be required because patients reach the maintenance dose more quickly with genotyping. Due to significant uncertainties regarding important assumptions in their decision-analytic Markov model, Verhoef et al. [13] stated that it was too early to decide whether Dutch patients with atrial fibrillation starting phenprocoumon should be genotyped, even though pharmacogenetic-guided dosing of phenprocoumon could improve health outcomes slightly in a cost-effective way. The main concerns were the effectiveness of a genome-guided dosing regimen and the costs of the genetic test. A subsequent study by Verhoef et al. [14] investigated the cost-effectiveness of a pharmacogenetic dosing algorithm for phenprocoumon and acenocoumarol compared to clinical dosing in the Netherlands. They found that genome-guided dosing increased costs for acenocoumarol treatment by €33 and quality-adjusted life-years (QALYs) by 0.001 and as such improve health only slightly when compared with clinical dosing. The incremental cost-effectiveness ratio was €28,349 per QALY gained for phenprocoumon and €24,427 for acenocoumarol. Even though, at a willingness-to-pay threshold of €20,000 per QALY, the pharmacogenetic dosing algorithm was not likely to be cost-effective compared with the clinical dosing algorithm, the authors suggested that the availability of low-cost genotyping would make it a cost-effective option. For typical patients with nonvalvular atrial fibrillation, warfarin-related genotyping was unlikely to be cost-effective, yet might be cost-effective in those patients at high hemorrhagic risk. However, *CYP2C9*- and *VKORC1*-guided warfarin treatment was shown to be cost-effective in the UK and Sweden healthcare sectors with 6702 GBP and 253,848 SEK per QALY gained, respectively [15]. Of particular interest is the study of Jowett et al. [16] that focused on the time and travel costs that patients incur to themselves to attend anticoagulation clinics. These costs are potentially important because therapy success requires regular monitoring and, frequently, dose adjustment. In this study, patients incurred considerable costs when visiting anticoagulation clinics, and these costs varied by country, ranging from €6.90 (France) to €20.50 (Portugal) per visit. A broad economic perspective is clearly important when considering the cost-effectiveness of genome-guided warfarin treatment [17].

6.3 Examples of economic evaluations of personalized medicine interventions in medium- and low-income countries

At present, there is growing evidence of the cost-effectiveness of genome-guided drug treatment modalities in medium- and low-income countries worldwide that significantly impacts the pricing and reimbursement of these interventions. In the following paragraphs we provide key examples of economic evaluations of personalized medicine interventions in medium- and low-income countries, while a comprehensive list is provided in Simeonidis et al. [18].

6.3.1 Anticoagulation and antiplatelet therapies

Warfarin is one of the most studied drugs in Europe and one of the few for which data on cost-effectiveness is available worldwide and thus it is a useful case study to consider the potential cost-effectiveness of personalized medicine interventions outside high-income countries. In a recent economic evaluation that considered pharmacogenomics-guided warfarin treatment in elderly atrial fibrillation patients in Croatia, a country with low levels of integration of genomic interventions and relatively poor genomic literacy, it was shown that 97% of elderly patients with atrial fibrillation in the pharmacogenomics-guided group did not have any major complications, compared to 89% in the control group, and that the incremental cost-effectiveness ratio for pharmacogenomics-guided treatment compared to non-pharmacogenomics-guided treatment was €31,225 per QALY gained [19]. These results suggest pharmacogenomics-guided warfarin treatment is a cost-effective therapy option for this patient group. This result may also hold for the same and other anticoagulation treatment modalities in neighboring countries.

A second economic analysis assessed whether *CYP2C19*-guided genotyping was cost-effective for myocardial infarction patients receiving clopidogrel treatment in the Serbian population compared to non-genotype-guided treatment. This study reported that clopidogrel treatment coupled with

CYP2C19-guided genotyping may be cost saving for myocardial infarction patients undergoing primary percutaneous coronary intervention in Serbia [20]. In particular, it was shown that 59.3% of the *CYP2C19**1/*1 patients had a minor or major bleeding event versus 42.85% of the *CYP2C19**1/*2 and *2/*2 patients, while a reinfarction event occurred only in 2.3% of the *CYP21C9**1/*1 patients with a mean cost of €2547 compared to 11.2% of the *CYP2C19**1/*2 and *CYP2C19**2/*2 patients and a mean cost of €2799. Also, based on the overall *CYP2C19**1/*2 genotype frequencies in the Serbian population, a break-even analysis indicated that performing the genetic test prior to drug prescription was cost saving, saving €13 per person on average.

6.3.2 Antidepressants

The use of pharmacogenomic tests to guide treatment in major depression cases is another informative case study, and an economic evaluation undertaken by Olgiati et al. considered the relative cost-effectiveness of different treatments in high- and middle-income European countries [21]. The study described the cost utility of incorporating 5-HTTLPR genotyping prior to drug treatment in major depressive disorders. Two drugs, citalopram and bupropion, were evaluated based on their predicted response and tolerability. A model was constructed for European countries with high gross domestic product (GDP; Austria, Belgium, Cyprus, Czech Republic, Denmark, Finland, France, Germany, Greece, Ireland, Italy, Luxembourg, Malta, Netherlands, Portugal, Slovenia, Sweden, United Kingdom), middle GDP (Bulgaria, Poland, Romania, Slovakia), and middle-high GDP (Estonia, Hungary, Latvia, Lithuania). The results showed that overall treatment costs were higher in Eastern Europe than in Western Europe, as patients were more frequently treated in inpatient facilities in these locations although the service costs were actually lower. However, the cost-effectiveness of the genome-guided approach was the same in all countries ($1147 per quality-adjusted life-week (QALW) in countries with high GDP; $1158 per QALW in countries in the middle GDP group, and $1179 per QALW in countries in the middle-high GDP group). The authors suggest that the critical factor that determines whether the pharmacogenomic test is cost-effective is the cost-effectiveness threshold, which is proportional to the economical level of the county. The World Health Organization indicates an intervention is highly cost-effective in a particular country if it is cost-effective and has an ICER that is inferior or equal to the GDP per capita of that country, or if the ICER is between one and three times the GDP per capita (www.who.int/choice/costs/CER_thresholds/en). Of note, since cost data were from 2009, the current economic situation in the countries that were evaluated might have evolved. These results were determined to be robust against variations in all parameters except for the cost of the genetic test, which produced the greatest changes in the ICERs. It was suggested that as long as genetic analysis is relatively expensive its applicability is limited to the richest areas in the Eurozone.

6.3.3 Allopurinol

The cost utility of *HLA-B*5801* testing before treatment with allopurinol has been evaluated in the Thai population from a societal perspective [22]. This modeling study compared a hypothetical pharmacogenetics treatment with conventional treatment, with carriers of the *HLA-B*5801* allele receiving an alternative drug (probenecid) and all others receiving allopurinol. Model parameters were either derived from the literature [23] or calculated by the study team. The cost-effectiveness of pharmacogenetics treatment was sensitive to the medical costs associated with gout management, the incidence of allopurinol-induced SJS/TEN, the probability of death with SJS/TEN, and the cost of genetic testing.

The authors conclude that genotype testing for the *HLA-B*5801* allele is potentially cost-effective only in countries with a large number of subjects at risk, such as countries in Southeast Asian or countries with large Han Chinese populations, as well as countries with a high prevalence of the *HLA-B*5801* allele. The prevalence of this allele was reported to be as high as 5.5%–15% in Thai and Han Chinese populations, but as low as 0.6% and 0.8% in Japanese and European populations. A Korean study that considered a hypothetical cohort of gout patients with chronic renal insufficiency reported similar results [24]. This study took a national health payer's perspective and concluded that allopurinol treatment based on *HLA-B*5801* genotyping could be the cost-effective alternative, improving health outcomes compared to conventional treatment.

6.3.4 Human papillomavirus testing

Human papillomavirus (HPV) testing has been evaluated in several settings outside high-income countries. Vijayaraghavan et al. developed a decision model to determine the cost-effectiveness of several cervical cancer screening strategies in South Africa [25]. These strategies included conventional cytology, cytology followed by HPV testing for triage of equivocal cytology, HPV testing, HPV testing followed by cytology for triage of HPV-positive women, and co-screening with cytology and HPV testing. The primary outcome measures included QALY gained, incremental cost-effectiveness ratios, and lifetime risk of cervical cancer. Results showed that screening once every 10 years reduced the lifetime risk of cervical cancer by 13%–52% depending on the screening strategy used. When strategies were compared incrementally, cytology with HPV triage was less expensive and more effective than screening using cytology alone. HPV testing with the use of cytology triage was a more effective strategy, costing an additional 42,121 Rand per QALY gained. HPV testing with colposcopy for HPV-positive women was the next most effective option. Simultaneous HPV testing and cytology co-screening was the most effective strategy, with an incremental cost of 25,414 rand per QALY gained. The authors therefore concluded that HPV testing for cervical cancer screening was a cost-effective strategy in South Africa.

A second study evaluated the cost-effectiveness of cytology and HPV DNA screening for women aged 25–65 years in Lebanon [26]. In Lebanon, there is no national organized cervical cancer (CC) screening program, so screening is limited to those who can pay out of pocket. Hence, this study evaluated the cost-effectiveness of varying screening coverage from 20% to 70% and varying screening frequency intervals from 1 to 5 years. The model was calibrated to epidemiological data from Lebanon, including CC incidence and HPV type distribution. The results showed that at 20% coverage, annual cytologic screening reduced lifetime CC risk by 14% and had an incremental cost-effectiveness ratio of I$80,670 per life-year saved (YLS), far exceeding Lebanon's GDP per capita (I$17,460), a commonly cited cost-effectiveness threshold. By comparison, increasing cytologic screening coverage to 50% and extending screening intervals to 3 and 5 years provided a greater CC reduction (26% and 21%, respectively) at lower costs, when compared to 20% coverage with annual screening. Screening every 5 years with HPV DNA testing at 50% coverage provided greater CC reductions than cytology at the same frequency (23%) and was cost-effective, assuming a cost of I$18 per HPV test administered (I$12,210/YLS). HPV DNA testing every 4 years at 50% coverage was also cost-effective at the same cost per test (I$16,340). Increasing coverage of annual cytology was not found to be cost-effective. The analysis concluded that the current practice of repeated cytology in a small percentage of women is inefficient, while increased coverage to 50% with extended screening intervals provides greater health benefits at a reasonable cost and thus can more equitably distribute health gains. Notably, novel HPV DNA strategies offer greater CC reductions and may be more cost-effective than cytology.

In a third study, the cost-effectiveness of different cervical screening strategies was estimated in Iran, a country with a low incidence rate of invasive cervical cancer (ICC) [27]. A Markov model was constructed with parameters extracted from primary data and current literature. Eleven strategies were compared, including Pap smear screening and HPV DNA testing plus Pap smear triaging, with various starting ages and screening intervals. Model outcomes included lifetime costs, life years gained, QALY, and incremental cost-effectiveness ratios. The authors found that compared to no screening the 11 strategies prevented between 26% and 64% of all mortalities. The cost-effective strategy was HPV screening, starting at the age of 35 years and repeated every 10 years, which cost $8875 per QALY gained compared with no screening. Screening at five-year intervals was found to be cost-effective based on GDP per capita in Iran.

6.4 A generic model as tool for measuring cost-effectiveness of personalized medicine interventions in medium- and low-income countries

Demand is increasing for cost-effectiveness evidence to support the use and reimbursement of personalized medicine technologies and interventions in medium- and low-income countries. However, the generation of sufficient economic evidence to inform healthcare decision-making is costly and the resources to perform these analyses are limited (see Sections 6.1 and 6.2). Given this, generic economic models are urgently required that can be efficiently adapted on a country-by-country basis. Such models could help decision-makers to generate timely information on value to guide adoption decisions for personalized medicine interventions that could yield substantial health improvements. This approach could be particularly beneficial for low- and middle-income countries with severely constrained resources for performing personalized medicine economic evaluations [9].

In response to this challenge, an eight-step generic economic model has been developed and applied to the use case of *HLA-B*15:02* genotyping to predict carbamazepine-induced cutaneous reactions, with a user-friendly decision-making tool relying on user-provided input values [28]. The eight steps are described here and in Fig. 6.1.

(1) **Team building**: Put together a generic model team with relevant stakeholders and expertise represented (e.g., clinical, pharmacogenomics, economics) to employ a consensus evidence peer-review process,
(2) **Selection of the original model(s)**: Select and adapt of at least one existing peer-review model to make it generalizable by completing a detailed review of the model structure, variables, and assumptions,
(3) **Evidence generation**: Complete evidence review and obtain expert opinion as needed to provide supporting documentation for the new generalizable assumptions and parameter values,
(4) **Development of the generic model**: Determine which model variables and assumptions will be default values only, input values only, or both default and input options,
(5) **Face validity**: Compile model documentation sufficient for the team to complete its review and for others to reproduce the model,

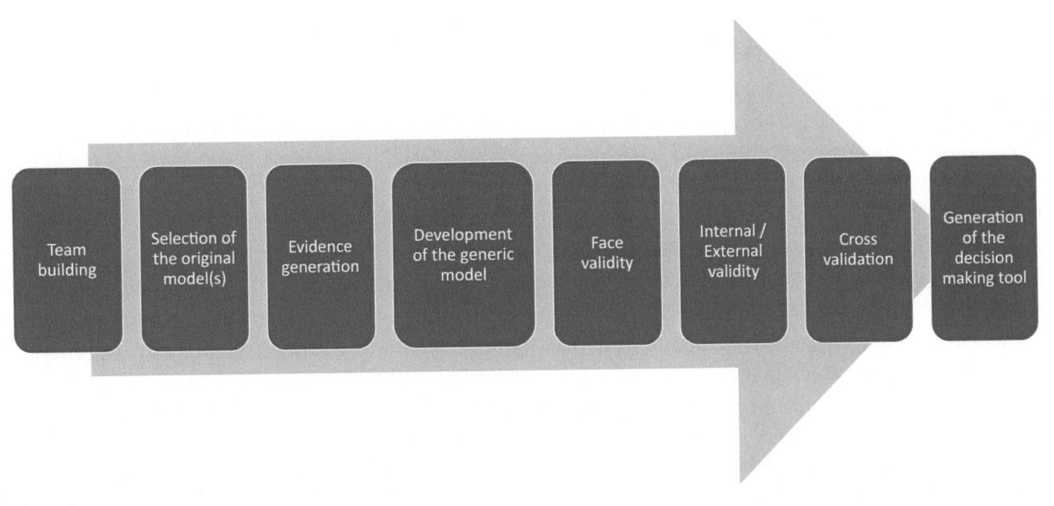

FIG. 6.1

Outline of the eight steps involved in the generic economic model developed to determine cost-effectiveness of *HLA-B*15:02* genotyping to predict carbamazepine-induced cutaneous reactions, which can be eventually expanded to other economic evaluations.

(6) Internal and external validity: Review, test, and revise the generic model by running base case and probabilistic sensitivity analysis simulations with independent reviewers experienced in model development,

(7) Cross validation: Validate the generic model using at least one specific model and multiple sets of input values, and

(8) Generation of the decision-making tool: Convert the decision analysis model to a user-friendly tool with instructions explaining the model, requirements for providing input values and results.

This generic model was transparently documented and validated and used to compare the cost-effectiveness results of three country-specific models for Singapore, Thailand, and Malaysia [28]. The generic model was successfully cross-validated using input values for six populations from these three countries and produced consistent results for *HLA-B*15:02* screening at country-specific cost-effectiveness threshold values.

This proof of concept demonstrates the feasibility of generic models to provide useful economic evidence on personalized medicine interventions, supporting their use as a pragmatic and timely approach to address a growing need. Focused implementation of generic models to generate value-based evidence for priority areas could therefore contribute to the timely realization of the benefits offered by personalized medicine.

6.5 Conclusion and future perspectives

The examples described in this chapter suggest that economic evaluations should be replicated in every country to inform implementation of pharmacogenomic-guided medical interventions, with decision models tailored to specific country characteristics. This is a daunting expectation given the rapidly increasing number of pharmacogenomics guidelines published by regulatory agencies in high income countries, such as the US FDA, and multi-national organizations, such as the Clinical Pharmacogenetics International Consortium. In addition, few medium- and low-income countries have the necessary resources and expertise to perform these analyses. The use of generic economic models could mitigate these challenges if these models would allow key variables, such as allele frequency and test and treatment costs, to vary by country. These models could be used by less experienced modelers with access to country-specific variables to generate a first approximation of cost-effectiveness, allowing different molecular genetic tests to be prioritized.

The application of population pharmacogenomics, namely, the impact of genetic ancestry in the pharmacogenomics implementation efforts, would contribute to making pharmacogenomics a useful tool that can be incorporated into clinical practice in every country. To reduce the disparity in practical use of pharmacogenomics in the developing world, projects such as the Euro-PGx project (which aims to determine the prevalence of pharmacogenomic biomarkers in different populations in Europe), could help to (1) enhance the understanding of pharmacogenomics in the developing world, (2) provide guidelines for medication prioritization, and (3) build infrastructure for future pharmacogenomic research studies. Although the Euro-PGx project is only focused on Europe, it provides a framework that could be replicated in other regions and countries. In addition, similar projects have been launched in other geographical regions, such as the 1000 Pharmacogenes project in Southeast Asia and the EmHEART project that aims to assess clinical utility and cost-effectiveness of genome-guided treatment modalities in cardiovascular disease patients in the United Arab Emirates. Other initiatives, such as the Human Heredity and Health in Africa (H3Africa) study, the Qatar Genome Project, and the Mexico National Institute of Genomic Medicine (INMEGEN) aim to build capacity and empower local researchers to spark a paradigm shift.

To conclude, it is crucial that research into the use of genetics in clinical practice is combined with economic analyses that determine the financial consequences of using these technologies to ensure that patients receive an acceptable level of care while effectively managing scarce healthcare resources. It is important that the scientific projects outlined in this chapter are continued to generate further insights into the efficient use of these healthcare technologies.

Acknowledgments

This work was endorsed by the Genomic Medicine Alliance Economic Evaluation in Genomic Medicine Working Group.

References

[1] Patrinos GP, editor. Applied genomics and public health. Burlington, CA, USA: Elsevier/Academic Press; 2020.

[2] Koutsilieri S, Tzioufa F, Sismanoglou DC, Patrinios GP. Unveiling the guidance heterogeneity of genome-informed drug treatment interventions among regulatory bodies and research consortia. Pharmacol Res 2020;153, 104590.

[3] Drummond MF, Sculpher MJ, Claxton K, Stoddart GL, Torrance GW. Methods for the economic evaluation of health care programmes. Oxford University Press; 2015.

[4] Fragoulakis V, Mitropoulou C, Williams MC, Patrinos GP. Economic evaluation in genomics medicine. Elsevier/Academic Press; 2015.
[5] Mitropoulou C, Katelidou D, van Schaik RH, Maniadakis N, Patrinos GP. Performance ratio based resource allocation decision-making in genomic medicine. OMICS 2017;21(2):67–73.
[6] Phillips KA, Douglas MP, Wordsworth S, Buchanan J, Marshall DA. Avaulability and funding of genomic sequencing globally. BMJ Glob Health 2021;6, e004415.
[7] Rebecchi A, Gola M, Kulkarni M, Lettieri E, Paoletti I, Capolongo S. Healthcare for all in emerging countries: a preliminary investigation of facilities in Kolkata, India. Ann Ist Super Sanita 2016;52:88–97.
[8] Zhang M, Wang W, Millar R, Li G, Yan F. Coping and compromise: a qualitative study of how primary health care providers respond to health reform in China. Hum Resour Health 2017;15:50.
[9] Snyder SR, Mitropoulou C, Patrinos GP, Williams MS. Economic evaluation of pharmacogenomics: a value-based approach to pragmatic decision making in the face of complexity. Public Health Genom 2014;17(5-6):256–64.
[10] Tangamornsuksan W, Chaiyakunapruk N, Somkrua R, Lohitnavy M, Tassaneeyakul W. Relationship between the HLA-B* 1502 allele and carbamazepine-induced Stevens-Johnson syndrome and toxic epidermal necrolysis: a systematic review and meta-analysis. JAMA Dermatol 2013;149:1025–32.
[11] Sukasem C, Katsila T, Tempark T, Patrinos GP, Chantratita W. Drug-induced stevens-johnson syndrome and toxic epidermal necrolysis call for optimum patient stratification and theranostics via pharmacogenomics. Annu Rev Genomics Hum Genet 2018;19:329–53.
[12] Mizzi C, Dalabira E, Kumuthini J, Dzimiri N, Balogh I, Başak N, et al. A european spectrum of pharmacogenomic biomarkers: implications for clinical pharmacogenomics. PLoS One 2016;11, e0162866.
[13] Verhoef TI, Redekop WK, Veenstra DL, Thariani R, Beltman PA, Van Schie RM, et al. Cost-effectiveness of pharmacogenetic-guided dosing of phenprocoumon in atrial fibrillation. Pharmacogenomics 2013;14:869–83.
[14] Verhoef TI, Redekop WK, De Boer A, Maitland-Van Der Zee AH, EU-PACT Group. Economic evaluation of a pharmacogenetic dosing algorithm for coumarin anticoagulants in The Netherlands. Pharmacogenomics 2015;16:101–14.
[15] Verhoef TI, Redekop WK, Langenskiold S, Kamali F, Wadelius M, Burnside G, Maitland-van der Zee AH, Hughes DA, Pirmohamed M. Cost-effectiveness of pharmacogenetic-guided dosing of warfarin in the United Kingdom and Sweden. Pharm J 2016;16:478–84.
[16] Jowett S, Bryan S, Mahé I, Brieger D, Carlsson J, Kartman B, et al. A multinational investigation of time and traveling costs in attending anticoagulation clinics. Value Health 2008;11:207–12.
[17] Verhoef TI, Redekop WK, Van Schie RM, Bayat S, Daly AK, Geitona M, et al. Cost-effectiveness of pharmacogenetics in anticoagulation: international differences in healthcare systems and costs. Pharmacogenomics 2012;13:1405–17.
[18] Simeonidis S, Koutsilieri S, Vozikis A, Cooper DN, Mitropoulou C, Patrinos GP. Application of economic evaluation to assess feasibility for reimbursement of genomic testing as part of personalized medicine interventions. Front Pharmacol 2019;10:830.
[19] Mitropoulou C, Fragoulakis V, Bozina N, Vozikis A, Supe S, Bozina T, et al. Economic evaluation of pharmacogenomic-guided warfarin treatment for elderly Croatian atrial fibrillation patients with ischemic stroke. Pharmacogenomics 2015;16:137–48.
[20] Mitropoulou C, Fragoulakis V, Rakicevic LB, Novkovic MM, Vozikis A, Matic DM, Antonijevic NM, Radojkovic DP, van Schaik RH, Patrinos GP. Economic analysis of pharmacogenomic-guided clopidogrel treatment in Serbian patients with myocardial infarction undergoing primary percutaneous coronary intervention. Pharmacogenomics 2016;17:1775–84.
[21] Olgiati P, Bajo E, Bigelli M, De Ronchi D, Serretti A. Should pharmacogenetics be incorporated in major depression treatment? Economic evaluation in high-and middle-income European countries. Prog Neuro-Psychopharmacol Biol Psychiatry 2012;36:147–54.

[22] Saokaew S, Tassaneeyakul W, Maenthaisong R, Chaiyakunapruk N. Cost-effectiveness analysis of HLA-B* 5801 testing in preventing allopurinol-induced SJS/TEN in Thai population. PLoS One 2014;9, e94294.
[23] Somkrua R, Eickman EE, Saokaew S, Lohitnavy M, Chaiyakunapruk N. Association of HLA-B* 5801 allele and allopurinol-induced Stevens Johnson syndrome and toxic epidermal necrolysis: a systematic review and meta-analysis. BMC Med Genet 2011;12:118.
[24] Park DJ, Kang JH, Lee JW, Lee KE, Wen L, Kim TJ, et al. Cost-effectiveness analysis of HLA-B5801 genotyping in the treatment of gout patients with chronic renal insufficiency in Korea. Arthritis Care Res 2015;67:280–7.
[25] Vijayaraghavan A, Efrusy M, Lindeque G, Dreyer G, Santas C. Cost effectiveness of high-risk HPV DNA testing for cervical cancer screening in South Africa. Gynecol Oncol 2009;112:377–83.
[26] Sharma M, Seoud M, Kim JJ. Cost-effectiveness of increasing cervical cancer screening coverage in the Middle East: an example from Lebanon. Vaccine 2017;35:564–9.
[27] Nahvijou A, Daroudi R, Tahmasebi M, Hashemi FA, Hemami MR, Sari AA, et al. Cost-effectiveness of different cervical screening strategies in Islamic Republic of Iran: a middle-income country with a low incidence rate of cervical cancer. PLoS One 2016;11, e0156705.
[28] Snyder SR, Hao J, Cavallari LH, Geng Z, Elsey A, Johnson JA, Mohamed Z, Chaiyakunapruk N, Chong HY, Dahlui M, Shabaruddin FH, Patrinos GP, Mitropoulou C, Williams MS. Generic cost-effectiveness models: a proof of concept of a tool for informed decision-making for public health precision medicine. Public Health Genom 2018;21(5-6):217–27.

CHAPTER

Theoretical models for economic evaluation in genomic and personalized medicine

7

Vasileios Fragoulakis[a], George P. Patrinos[b,c,d], and Christina Mitropoulou[a,c]

[a]The Golden Helix Foundation, London, United Kingdom, [b]Department of Pharmacy, School of Health Sciences, University of Patras, Patras, Greece, [c]Department of Genetics and Genomics, College of Medicine and Health Sciences, United Arab Emirates University, Al-Ain, United Arab Emirates, [d]Zayed Center of Health Sciences, United Arab Emirates University, Al-Ain, United Arab Emirates

7.1 Introduction

When conducting economic evaluations of health services, we often seek to identify the available options (or "health technologies") that maximize social welfare in a country or a healthcare system [1]. To make that selection, the political authority responsible will first have to precisely define all the available options and determine the associated costs and the benefit they provide for all relevant medical conditions [2]. This is done by expressing the cost of a health technology in a single monetary unit (e.g., euros or dollars) while expressing the benefits either through a specific evaluation or in a generic and comparable metric of evaluation independent from the specific medical conditions. In the case of a cancer patient undergoing a specific treatment, for example, we might calculate the monetary cost of the treatment, including drugs, hospitalization, treatment of side effects, and laboratory tests. The benefit for that patient could be defined in terms of "survival" or total "quality-adjusted life-years" (QALYs). [3] It could also be defined in terms of "progression-free survival" (PFS), which is a clinical measure of effectiveness that indicates the period of time over which the patient has remained stable with regard to disease progression, based on whether or not the tumor has shrunk, or in terms of "time to progression" (TTP), that is, the length of time from the date of diagnosis or the start of treatment until the disease starts to worsen or spread to other parts of the body. In clinical terms, measuring the TTP is one way to see how well a new treatment works. Measures of effectiveness such as PFS or TTP are clinical and are associated with specific conditions [4]. Conversely, QALY is a generic effectiveness measure that is not used for specific conditions only and might be used as a comparator for health technologies in different fields.

In cardiology, for a hypertensive cardiology patient, the costs would include drugs, tests, treatment of side effects, and hospital admissions. Here, too, the benefit can be defined in terms of the overall survival or QALYs, "percent reduction in pressure," or other equivalent measures proposed by clinical scientists in this specialty. Obviously, to be able to compare our cancer patient with our cardiology

Economic Evaluation in Genomic and Precision Medicine. https://doi.org/10.1016/B978-0-12-813382-8.00007-0

patient, we can only compare benefits expressed in terms of a "common denominator," such as the QALY, which includes both the concept of quality and the concept of quantity.

The primary question that health economics attempts to answer is how to distribute the available funds to the various public and private healthcare providers (hospitals, healthcare organizations, pharmaceutical companies, etc.) to cover the existing needs as much as possible and to help the greatest possible number of people in an economically viable way. At the technical level, when conducting economic evaluations that attempt to examine matters of resource distribution (e.g., how much money from the total healthcare budget should be allocated to cardiac patients, cancer patients, primary care patients, etc.), we use common effectiveness measures such as the QALY.

In practice, when evaluating an innovative treatment in comparison to an existing treatment, we determine the incremental cost-effectiveness ratio (ICER) [5]. This ratio is the difference between the overall cost of the two health technologies divided by the difference in benefit. The ICER indicates the additional number of resources that must be expended in order to provide one additional year of life to society, based on current medical knowledge. If, for example, the cost of a new treatment is 20,000 euros and the benefit is 10 QALYs, and the cost of the standard treatment is 10,000 euros and the benefit is 5 QALYs, the ICER is as follows:

$$\text{ICER} = (20,000 - 10,000)/(10 - 5) = €2000/\text{QALY}$$

The manager of the healthcare system resources may choose to reimburse the new treatment by paying 2000 euros for each additional unit of benefit. If this cost is considered reasonable and affordable, then the new treatment will be adopted. If this cost is considered excessive by the decision makers, then the new treatment will not be reimbursed by the state or there will be pressure to reduce the price of the new technology to acceptable levels.

Now let's examine a different case. If the cost of a new treatment is 10,000 euros and the benefit is 5 QALYs, and the cost of the standard treatment is 20,000 euros and the benefit is 10 QALYs, the ICER is as follows:

$$\text{ICER} = (15,000 - 20,000)/(5 - 10) = €1000/\text{QALY}$$

In this case, the new treatment is cheaper but provides lower benefit. This is not unheard of since healthcare systems seek to reduce costs by introducing cheaper generic drugs. The ratio here is interpreted in the opposite way: it indicates the amount of resources that can be saved if society agrees to "scarify" one year of life compared to the standard treatment used for a specific condition.

In most cases, the standard treatment is cheaper and less effective, whereas the new treatment is more expensive and effective, as efforts are made to improve effectiveness over time. So, to reach a final decision as to which of the two health technologies should be adopted by a country's healthcare system, we should have a rule to determine whether the ICER is attractive or not. Indeed, we need to compare the amount of money **needed** to achieve greater effectiveness (the ICER) with the amount of money that the responsible agencies (budget holders) are **willing** to invest to obtain it. The latter amount is called "willingness to pay" and is denoted by "wtp" or by the Greek letter λ, which represents the state's institutional representatives' willingness to invest additional resources (negative consequence) to obtain more QALYs (positive development) or to sacrifice QALYs (negative consequence) when specific requirements with regard to cost reductions are met (positive development). When the ICER is lower than the λ, then a new health technology is considered to be advantageous for the society and is adopted by the system. If, for example, λ=50,000 and ICER=30,000, the new treatment is

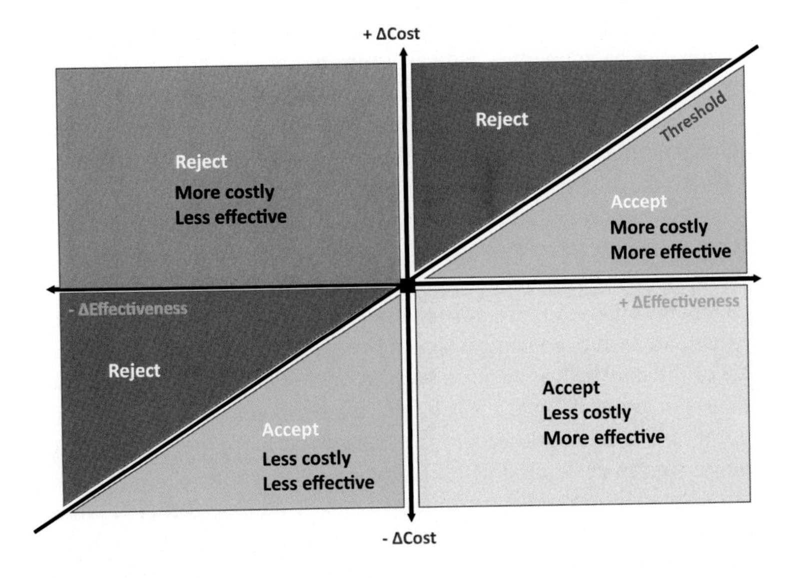

FIG. 7.1

The cost-effectiveness plane. "Δ" stands for "Difference"; the term "threshold" indicates willingness to pay (wtp) or lambda (λ).

advantageous because each additional QALY demands 30,000 euros and society is willing to pay up to 50,000 euros. Fig. 7.1 presents the λ diagram.

We should note that calculating the ICER is a **technical matter** that requires measuring and comparing different health technologies, whereas determining the λ is a **societal choice** that reflects the ability of a healthcare system to allocate resources to adopt innovations. We should also stress that "societal preference" and "societal choices" are complex processes that may depend on the national culture, personal preferences, and the degree of social consensus with regard to "social fairness," and therefore cannot be examined only through economics. Health economics will only perform a technical analysis that will simplistically assume that societal preferences coincide fully with the preferences of the state's institutional representatives and can be expressed numerically. (For technical details see Fragoulakis et al [2].)

Even though this rule appears to be methodologically attractive (though simplistic), it has several drawbacks that need to be addressed [6–11]. For example, an option may seem cost-effective (another way of saying that it is advantageous for the society or the so-called value for money), but this does mean that the healthcare system is able to finance its additional cost compared to the standard treatment. Cost-effectiveness indicates whether the ratio between cost and benefit is attractive but does not examine whether the funds required to cover the excess burden from a new treatment are available. For example, as we can see in Fig. 7.1, all the options that lie on the curve that defines the λ (which extends indefinitely) could be adopted, but this would require limitless resources, which is obviously not the case. In practice we use a different type of analysis called "budget impact analysis," which determines the actual effects on the available funds. Since there is no definite rule to correlate cost-effectiveness with budget impact, the adoption of a new technology usually depends also on the political, historical, and social circumstances of each country, so it is arbitrary.

Thus, the major weaknesses of the analysis so far are that (1) it assumes that the available funds are limitless and can be directly adjusted to the cost of new technologies, (2) λ is determined arbitrarily, (3) the budget analysis does not depend on λ, something that is not true for any healthcare system, (4) innovation, as defined by the difference in effectiveness between health technologies (difference in QALYs), is not included in any way in the model but is assumed to be socially irrelevant, and (5) issues of ethical patient management can be examined using purely economic criteria.

For genomic medicine, the development of such a model is important because the newer personalized medicine treatments appear to be significantly more advantageous compared to older standard treatments [12]. If these are evaluated using a model that does not take into account or reward the level of innovation, years of life may be lost because such technologies may be rejected, which will actually impart significant costs in terms of society's health. Their economic cost, on the other hand, is usually greater than the cost of older treatments because they require certain expensive genetic tests; therefore a clear, objective metric for the evaluation of all technologies is needed.

For instance, an economic evaluation in genomic medicine assessed [13] whether pharmacogenomic (PGx)-guided warfarin treatment of elderly ischemic stroke patients with atrial fibrillation in Croatia is cost-effective compared with non-PGx therapy. Primary analysis indicated that 97.07% (95% CI: 94.08%–99.34%) of patients belonging to the PGx-guided group did not develop any major complications compared with the control group (89.12%; 95% CI: 84.00%–93.87%, $P<0.05$). The total cost per patient was estimated at €538.7 (95% CI: €526.3–551.2) for the PGx-guided group versus €219.7 (95% CI: €137.9–304.2) for the control group. In terms of QALYs gained, total QALYs were estimated at 0.954 (95% CI: 0.943–0.964) and 0.944 (95% CI: 0.931–0.956) for the PGx-guided and the control groups, respectively. The true difference in QALYs was estimated at 0.01 (95% CI: 0.005–0.015) in favor of the PGx-guided group. The ICER of the PGx-guided versus the control groups was estimated at €31,225/QALY. Overall, the data indicated that PGx-guided warfarin treatment may represent a cost-effective therapy option for the management of elderly patients with atrial fibrillation who developed ischemic stroke in Croatia.

In such cases, it is important to evaluate innovation in terms of reduced bleedings and increased QALYs to investigate how λ correlates with the actual effects on available funds and to develop a new type of economic thinking that relaxes the restrictive assumption that the innovative nature of a health technology has no effect on the proportional willingness to pay.

To resolve all these issues so that such an analysis can be applied in genetics, a new methodological model for the evaluation of health technologies has been developed. This model is called the Genome Economics Model (GEM) and we describe it in the following section.

7.2 The genome economics model

The GEM [14] proposes that at least two limits should be applied in relation to the λ. One of these limits is defined by the budget and determines which options are cost-effective as well as cost-affordable, meaning that they can be adopted by the healthcare system based on the available funds. Options that do not meet the criterion of cost-effectiveness (ICER$<\lambda$) are rejected, but also any option that is cost-effective ($\lambda>$ICER) but does not comply with the budgetary criterion because it exceeds the maximum amount of resources that may be allocated for a new treatment is also rejected. The GEM also applies another limit, which we analyze later. According to the previous analysis, the budget holder may accept a less effective health technology to reduce costs, but it is obvious that this process cannot continue indefinitely. As mentioned previously, the λ may be used to identify health technologies that can economize on resources by sacrificing effectiveness. The rationale here is that the resources saved will be redistributed to some other sector of the healthcare system and therefore will increase social welfare overall. This argument is reasonable, but it is not entirely acceptable. Obviously, in times of economic depression, an economy can limit healthcare expenditures and consequently the average level of the services provided, but only up to a certain point. The practice of medicine involves several important moral issues that will not permit the adoption of options with much lower effectiveness than current practice based solely on economic criteria. If we follow this example to its extremes to demonstrate its methodological inconsistency, let us consider a society that ceases to provide health care to certain of its citizens in order to save resources. This would certainly reduce healthcare expenditures, but it would blatantly undermine the constitutional acquits and would irreversibly damage social cohesion. Therefore, the GEM assumes that there is also a limit for health technologies that can save resources. The amount of resources a society is willing to save sacrificing the effectiveness, is not necessarily the same with the amount which is willing to pay for one unit of effectiveness. Relevant studies have shown that the amount of money that a society demands in order to sacrifice one already obtained QALY is at least twice the amount it is willing to invest in order to achieve one additional QALY [15]. In other words, once a society has achieved a certain level of health through technology, it is reluctant to sacrifice that effectiveness based solely on economic criteria. Despite this, the matter of the limit had not been discussed until now, at least theoretically, as presented by the GEM. Fig. 7.2 presents a graphical explanation of all the aforementioned arguments.

In addition, the GEM addresses another, perhaps more important, point. According to the preceding analysis, λ is determined by the society and can differ between countries or between conditions, but it is *fixed* for each specific condition. In the United Kingdom, for example, λ is usually set at 25,000–30,000 GBP, while other suitable amounts proposed by the World Health Organization are 50,000 EUR/QALY or three times the per capita national income. All these proposed values are significant and constitute guides to help societies decide which health technologies are cost-effective and which are not. We should specify, however, that such a "fixed-λ" approach assumes that the willingness to pay does not consider the magnitude of the difference between the current treatment and an innovative new treatment. If, for example, the difference in effectiveness between two treatments was two QALYs and the λ was 50,000 euros, then the total available funds that the budget holders would be willing to invest would be $2\times50{,}000$; for an arbitrary difference of a QALYs, the willingness to pay would be $a\times50{,}000$. This means that, in the preceding analysis, the amount of 50,000 is always fixed and does not depend on a; this is a simplistic assumption that is not consistent with the concept of utility, as used in economics. In economics, the utility of additional consumption is at first proportionally greater and gradually decreases. To incorporate this concept in the GEM as well, the λ was modified to correlate with the magnitude of the difference between the standard treatment and the new treatment (a). Fig. 7.3 describes this new approach for determining the λ. The curve is S-shaped to incorporate all situations that could arise when introducing a new technology into the healthcare system. If, for example, a new technology is minimally better than the standard treatment in terms of effectiveness (and similar in terms of manufacture, active ingredient, etc.), we would expect to see a very small λ because the budget holders would consider the two technologies almost identical based on evidence from the final

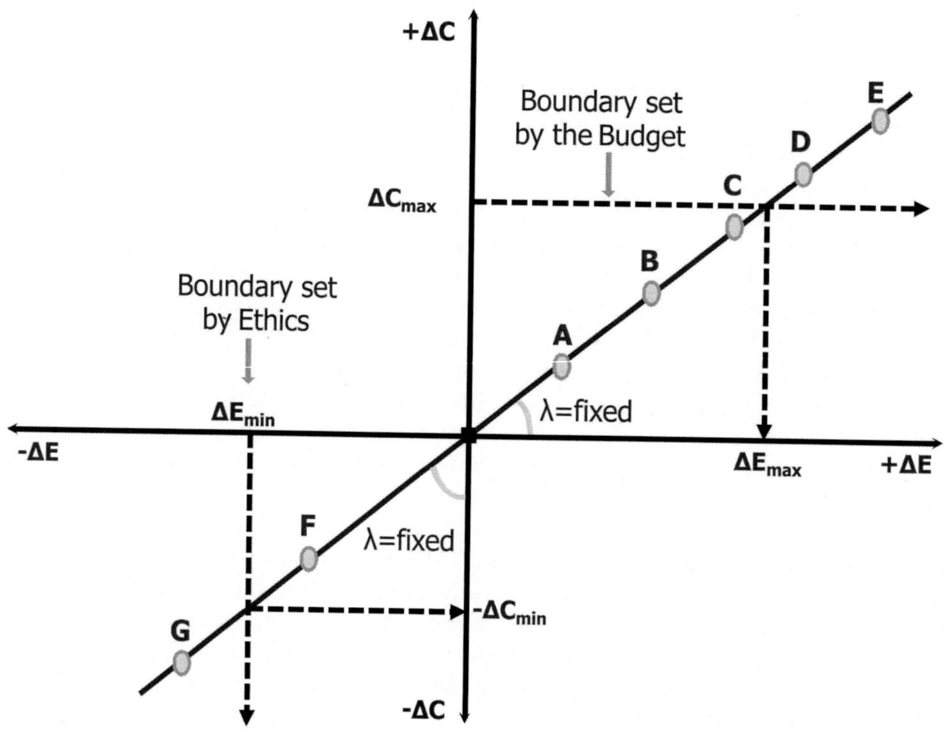

FIG. 7.2

Cost-Effectiveness Plane in GEM. The basis of the proposed Genome Economics Model (GEM). The classical cost-effectiveness plane is depicted. Various health interventions are indicated as A–G, lie on the straight line, and can be all adopted and reimbursed by the national healthcare systems in accordance with the classic model, which assumes unlimited budget availability. According to GEM, interventions such as "A" or "B" represent "average cases" with increased ΔE and reciprocally higher costs, which together with "innovative" technologies such as "C" (e.g., those described in pharmacogenomics and genomic medicine with a high ΔE) could be reimbursed by national healthcare systems. However, according to GEM, interventions with a much higher ΔE, such as "D" and "E," also bear higher costs, which make them unaffordable for adoption in a real-life situation where unlimited budget availability is never the case. Similarly, interventions such as "G" represent a technology with a cost-saving profile but also less effectiveness, which again make it ineligible for adoption.

Adapted from Fragoulakis V, Mitropoulou C, van Schaik RH, Maniadakis N, Patrinos GP. An alternative methodological approach for cost-effectiveness analysis and decision making in genomic medicine. OMICS 2016;20(5):274–282.

outcomes. If the new technology had a small difference in effectiveness, then the λ would be lower than the one proposed by the classic model, with a tendency to increase, and if the new technology had a significant difference to the standard treatment, then the λ would be greater than the one proposed by the classic model. As we approach the income restriction, of course, the additional amount of money that society is willing to invest to obtain a little more effectiveness would tend to become zero, since the value of money at that point would be more important. The highest effectiveness that the society is determined to incrementally reimburse should be specified in advance in this model.

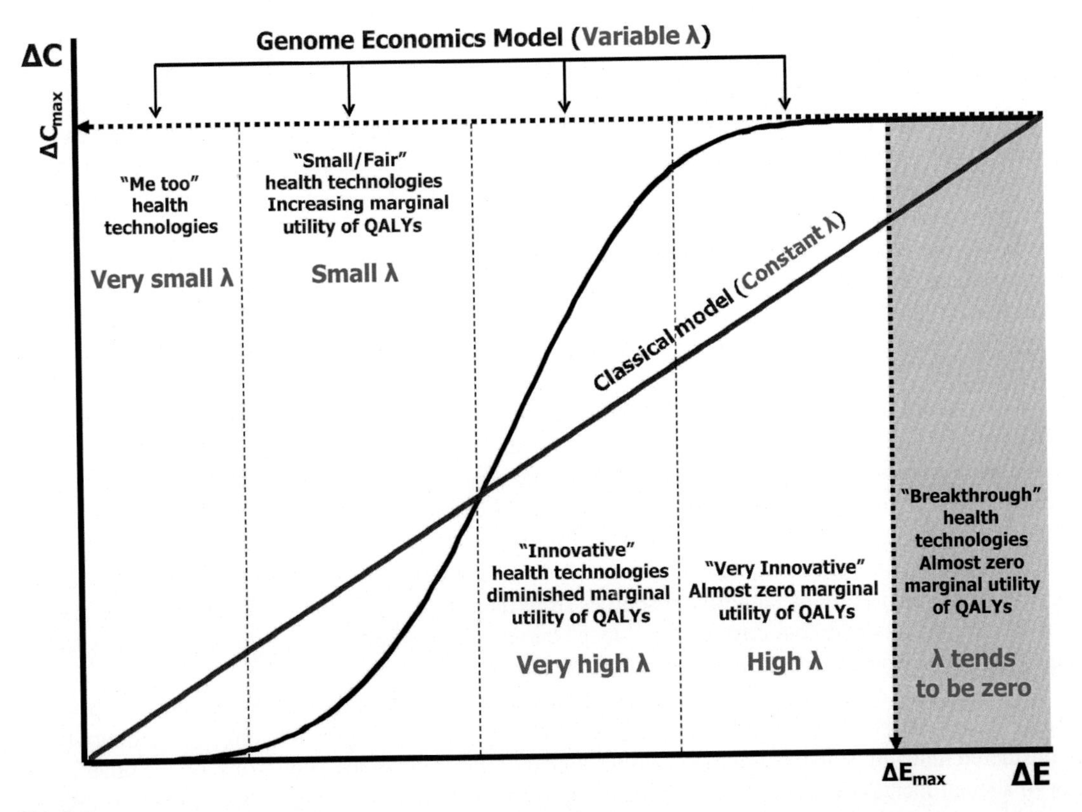

FIG. 7.3

Cost-Effectiveness Plane in the GEM versus the Classic Model. In the classic model, λ is constant, depicted as a straight line. However, in GEM, λ is variable, spanning from technologies with a very small innovation (me too) with a reciprocal small average λ to very innovative technologies with a high average λ. ΔE=difference in the effectiveness between the standard and the new intervention, depicted in the horizontal axis; ΔC=the difference in the total cost between the standard and new intervention per patient, depicted in the vertical axis; ΔC_{max}=the maximum available budget (per patient) is willing to afford the budget holder to capture the maximum expected effectiveness, which can be additionally reimbursed (=ΔE_{max}); "marginal utility" refers to the additional satisfaction the society gains from the production of one more unit of effectiveness.

Adapted from Fragoulakis V, Mitropoulou C, van Schaik RH, Maniadakis N, Patrinos GP. An alternative methodological approach for cost-effectiveness analysis and decision making in genomic medicine. OMICS 2016;20(5):274–282.

According to this analysis, innovation is an important determining factor to be considered in the reimbursement of health technologies. If a health technology is demonstrably different in **final health indicators** (indicators that translate to increased survival and quality), it should be reimbursed accordingly, always within the country's capabilities and based on its available funds. This does away with the indifferent proportional reimbursement and resources are transferred from undistinguished providers to those that have actually made a significant additional contribution.

It should be noted that the concept of innovation is very important, both in the health sciences and in other fields [16]. In many cases, innovation will not be readily understood by non-experts who do not have the necessary background to appreciate matters of production processes, quality of raw materials, safety, precision, and so on. This certainly includes health economists who are the ones that usually conduct the economic evaluation of health technologies since their specialty concerns a specific technical analysis and does not extend to purely clinical work. Thus, GEM and its analyses will define the results only narrowly and with obvious final indicators that can be measured objectively at this time. In this sense, the concept of innovation could be extended to include issues relating to safety, ease of use, improvement of the manufacturing conditions and methods that promise to show a difference in the final health indicators in the future, which has not been achieved yet, and more. In all such cases, a quantification of the innovation should be attempted to allow a comparison between the different health technologies.

Also, according to the GEM, the maximum amount of money that can be paid to achieve maximum effectiveness is predetermined; this is also a significant difference to the classic analysis. In that sense, the model assumes that the society and the relevant scientific associations determine in advance what is considered to be innovation and how much this should be reimbursed in the near future. The classic model is now an individual case of all possible forms of λ. Of course, the exact shape of the sigmoid curve might be different in various societies to indicate the degree to which each society rewards innovation by transferring presumptive resources from one healthcare provider to another. This reward for innovation is not without cost; a very "steep" curve would essentially reward only certain capable candidates but would set significant restraints on joint enterprises.

In developing countries, where resources are limited and therefore the value of money is especially high, there must be a significant difference between health technologies for a new technology to be considered innovative and, as such, to be reimbursed. In more developed countries, on the other hand, even a small increase in benefit could be enough for budget holders to invest more so that healthcare providers will adopt it, since the available funds are greater in those countries.

According to the GEM, these curves that define innovation are not expected to remain constant for different conditions, in different countries, or even over time. Obviously, the introduction of new health technologies formulates new standards of public attitude and behavior; the process of technology assessment and evaluation is continuous and much more dynamic than it appears to be at first glance. This model, which we explain, is a first attempt in conducting an analysis that includes the concept of innovation and the concept of societal preferences and clinical/economic data in a single analysis.

7.3 Generalization of the genome economics model

Although the GEM addresses certain simplifying assumptions already mentioned, in the form previously presented it cannot resolve the problem of distribution of a given budget to different health technologies that treat different conditions to achieve maximal societal utility. Specifically, the budget for each condition was considered to be **given** and **externally** determined, an assumption that may be partially valid and might reflect reality to some extent. Indeed, in many cases the management of healthcare resources does not follow any maximization models but is based on political and historical circumstances. However, it is important to attempt to develop a model aimed at maximizing utility, as described previously.

To do this, the GEM was generalized to create a model (the "gGEM") that can suggest a solution to maximize utility for all medical conditions. An advantage of this model is the fact that it does not require that the effectiveness of different conditions be assessed with the same metric (the QALY), although this would not be prohibitive [17]. Effectiveness is measured with a "performance ratio" expressed in percentage units. This model describes most closely the usual case in medical practice, as it assumes that a new health technology would probably have greater effectiveness and higher cost; this corresponds only to the upper right quadrant in the cost-effectiveness plane (Fig. 7.4).

Therefore, the cases that involve health technologies with lower overall cost and lower benefit than the standard treatment are not described in the model. It should be noted that in developed countries the share of healthcare expenditures is large and healthcare systems are under pressure to limit their expenses. To do this, prices are frozen, prescriptions are limited, and volume/price agreements are made to curb the rate of increase or even to reduce expenditures. In emerging countries, however (China, Russia, Brazil, etc.), the relevant expenditures are expected to increase; for this reason, the model only describes the upper-right quadrant of the cost-effectiveness diagram to address those countries

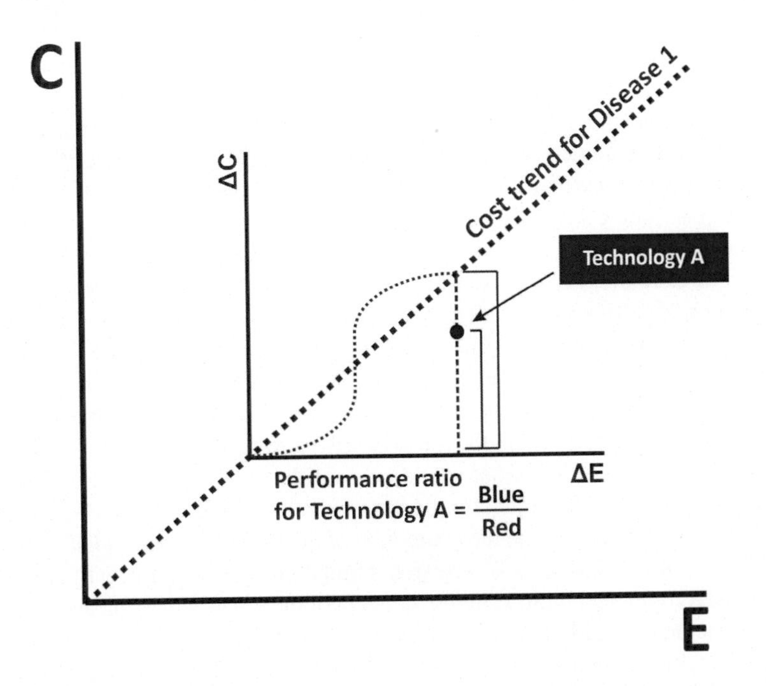

FIG. 7.4

Definition of Performance Ratio, which determines the relative performance of a healthcare technology based on a relative basis. For instance, a 20% performance indicates that the cost of a technology is only 80% (1%–20%) of the total cost that the insurance funds are willing to invest to gain the specific effectiveness this technology offers. When performance ratio equals 100%, this indicates that the new technology is offered at the same cost with the standard technology and thus represents a dominant choice. A negative performance ratio indicates technologies that must be rejected by the healthcare system.

Adapted from Fragoulakis V, Mitropoulou C, Katelidou D, van Schaik RH, Maniadakis N, Patrinos GP. Performance ratio based resource allocation decision-making in genomic medicine. OMICS 2017;21(2):67–73.

primarily. Nevertheless, we should note that the gGEM can be expanded to the third quadrant to include other countries without significant amendments to the substance of the analysis.

The gGEM is developed in six steps, as described next.

(a) Initially (step *a*), all possible treatment options for all different conditions must be described. This step effectively requires that all available options be included: the "most expensive" treatments, the "cheapest" treatments, the "do nothing" option, standard and alternative approaches, and so on. This is an important step because, in most cases, the available research data are supplied by healthcare providers who have a vested interest in entering a healthcare market and possess the necessary resources to conduct expensive clinical trials, economic analyses, and more. This is why most analyses and publications involve technologies with a large share in the relevant market. In this way, some social welfare may be lost because certain available options have not been assessed or have not yet been approved for use in the general population due to economic or other limiting factors.

(b) The second step (step *b*) requires that all these therapeutic approaches be expressed in a single monetary unit. Their cost could be calculated from the literature or with the standard procedures of the economic evaluation for costs, with the appropriate conversions where necessary.

(c) The third step (step *c*) consists of selecting the effectiveness metric to be used for each individual condition. As mentioned, to conduct a comparison using the classic models, all treatment options must be expressed in a single unit, usually the QALY. With the gGEM, this is not necessary. The model's only requirement is that the effectiveness metric be one-dimensional and can be expressed in an explicit quantitative way. It should be noted that the selection of any effectiveness criterion must be done with care since the conversion of certain metrics to others may entail significant uncertainty. For example, PFS is a significant clinical indicator for cancer patients because it indicates the period of time over which the patient remains stable and their disease does not progress; however, there isn't necessarily a direct correlation between PFS and survival. Patients with different PFS may have similar survival and therefore the indiscriminate use of this metric may lead to incorrect results. This step is concluded when the benefit and the cost of all treatment options have been described. Then, the various scientific groups involved in the treatment of different conditions must define what would be considered "maximum innovation" in the near future, in terms of the selected indicator. For example, if the PFS is selected for a type of cancer, maximum innovation would define the magnitude of the PFS that would satisfy such a designation (e.g., "12 months after three years of research, compared to 10 months with current standards"). Please note that a compound multi-criterion concept of the effectiveness measure could be created, which would include the additional patient survival, the effect on quality of life, the ease of use of the particular health technology, whether its form reduces the chance of misuse or dosage errors, whether or not is has fewer side effects, the side effect profile, and so on. Other factors that could be considered are the therapeutic properties and mode of action of a treatment, whether or not this treatment meets patient needs, whether it provides multiple options to the attending physician, and whether, and to what extent, it can supplant other forms of treatment. To produce such a metric, all relevant factors must be included and weighted appropriately so as to provide a final assessment of any therapeutic approach.

(d) The next step (step *d*) is to determine the mean cost for each unit of effectiveness by dividing the overall cost by the overall effectiveness. Then, linear prediction is used to estimate the cost that corresponds to the effectiveness at the point of maximum innovation mentioned previously. In this approach, any relative increase in the cost of a technology must be associated with a corresponding relative increase in benefit. If, for example, a certain health technology has a cost of 20,000 and a benefit of 10 QALYs, then a cost increase to 30,000 (50%) should provide at least a 50% increase in QALYs, to 15, to be economically acceptable as a reimbursable new treatment. We should stress two points here. First, this rationale is not fully accepted by the scientific community and many health economists realize that specific institutional practices need to be amended [18]. Second, such a reimbursement rule prevents the exponential increase of the cost of health technologies, a worrying event that jeopardizes the future viability of healthcare systems. Indeed, new health technologies today usually offer increased effectiveness but also have proportionally much higher purchasing cost. This is associated with the cost of development of each technology as well as the providers' efforts to maximize their profits as for-profit businesses operating in a difficult and competitive environment. Returning to our model, we must note that the mean cost per unit of effectiveness is expected to be different for different conditions, which is not surprising since both costs and benefits follow different patterns each time. For example, it is harder to obtain one additional unit of effectiveness for a cancer patient than for a cardiac patient and thus the relative cost is expected to be different in these two patient groups.

(e) The fifth step (step *e*) in the gGEM approach is to input a sigmoid curve that defines the willingness to pay differently depending on the level of innovation. As mentioned, patients and budget holders will perceive differently a treatment that provides a large difference in effectiveness (a truly innovative treatment) and a treatment with minimal contribution. In those cases, they will provide (proportionately) different reimbursement for additional units of effectiveness. The approach used here rewards healthcare providers that create innovation and "punishes" those that cannot work innovatively in an economically viable way.

(f) The sixth and last step (step *f*) of the model defines the concept of "performance ratio" (PR) for an available technology. The PR is defined as "1 minus the ratio of marginal cost of the technology divided by λ at the corresponding point of effectiveness." A negative PR indicates that policymakers are not willing to adopt that technology because the cost exceeds the maximum amount they are willing to pay to obtain it and therefore the society will lose. In such a case, that technology can be reimbursed only by private payers that can and will pay that cost. The most favorable scenario for a specific technology is a PR approaching 1 because that is when there is the greatest difference between the cost of the technology and the amount that society would be willing to invest to obtain it. For a graphical explanation, see Fig. 7.4.

The rule used by the model is that the available resources will be allocated to those health technologies that show the greatest PR values until the budget is fully exhausted. This model assumes that policymakers wish to maximize the PR and not the overall QALYs, as usually happens. The concept of innovation is included twice in the gGEM; first, the model considers the extent to which a health technology has improved a specific health indicator that represents effectiveness, and second, innovation is correlated to the cost needed to achieve it.

In this way, all health technologies, regardless of whether they have a high or low marginal benefit or are equivalent to a standard treatment, can be compared to each other and proven to be innovative with regard to effectiveness or "innovative" with regard to production cost.

7.4 Perspectives

The establishment of a new system for the distribution of healthcare resources in the most beneficial way for a society is an important step that ensures satisfactory healthcare services for its citizens and provides processes to ensure that these are economically viable. Such a distribution will also improve the potential for viable growth in the future. In contrast to the GEM, which determines the equilibrium of the system for a single condition, the gGEM attempts to resolve the problem of distributing healthcare resources to the entire healthcare system, aiming at maximizing social wellbeing without assuming that the available budget for each condition is fixed.

Such models need to be constructed for all countries, but especially in the developing world, to allow appropriate healthcare planning and to cover their needs in the near future and for future generations.

These models have certain characteristics that should be discussed. One significant feature is that QALYs are not necessarily used as the basic or comparative measure of effectiveness when comparing different health technologies used for different conditions. The reason for that is that the QALY is characterized by certain drawbacks that hinder comparison between conditions. How can the management of a case requiring total hip arthroplasty be compared to the improvement of the quality of life of a patient with macular degeneration? The QALY was not used in this analysis to avoid such methodological problems. We should note, however, that this approach is not entirely safe either. According to the gGEM, any clinical indicator that concerns a specific condition may be used. In that case, however, the metrics used will need adaptation. For example, if we use an indicator that can be easily improved in the future but does not correlate well with survival or quality, we might get falsely positive results with certain health technologies for a specific condition. To avoid this, an interdisciplinary clinical group should be appointed to determine the metrics that are most representative of the technologies being evaluated. In this way, the gGEM maximizes a metric of relative performance (the performance ratio), which is common throughout the healthcare sector and abstractive, to reach a resource distribution model. This means that the model's results will not necessarily maximize the QALYs, but they will direct resources to healthcare providers that have the best relative performance, so that they can continue to be productive in the future.

Another feature of these models is that their results are presented in a deterministic way (i.e., as point estimates). Cost and benefit data for health technologies usually come from clinical trials or observational studies and thus they follow specific distributions [19]. In practice there are specific techniques to determine the uncertainty for these assessments. If those were used, however, this model would be much more complex, and its results might be even more ambiguous. In addition, for the gGEM to be able to reach specific conclusions, it is necessary to determine in full the curve that defines the willingness to pay for each point of additional effectiveness. This is also difficult in practice since the budget holder might have a sense of the way they perceive innovation, but this might differ from the fully mathematical formulation that the model requires. In practice, the challenge is to quantify in full all of society's moral imperatives and the political views of the responsible authorities. We should also mention that the linear prediction described in step *c* can be performed in different ways, so the final results can be modified in accordance with the selected method.

Lastly, the models require some form of mathematical programming [20] at the end, a fact that increases the complexity of its application.

It must be noted that the gGEM tries to analyze the decision-making process of the healthcare sector concerning genomic medicine in a more realistic manner compared to previous attempts. Genomic medicine is a prominent sector and innovation in this sector typically involves making an outlay or expenditure in the expectation of future technologies that are more efficacious and viable in economic terms. Hence, the present model describes, in a dynamic manner, the link between the willingness to pay threshold, the size of innovation, and the cost of available technologies toward the budget limit. In addition, the present model proposes a process for the determination of equilibrium of the whole market, which is important in order to investigate the share of resources that could be given to genomic and the non-genomic technologies. Indeed, the costs of new healthcare technologies (genomic and non-genomic) might be the subject of critical coverage and thus there is a need to be correlated with the benefits that these technologies provide. The modification/adaptation of the model in order to better describe the real-world situation is the goal of future research in this field.

More generally, the two models for the economic evaluation of health technologies can be used as guides for thought, but they can be difficult to follow blindly when deciding whether to reimburse a specific option. Mathematics has a clearly attractive conciseness and purity, which are desirable in social sciences. Unfortunately, in practice, the number of factors involved in the actual management of resources is too large, which makes quantitative methods inadequate for use as the unique criterion in the analysis. In addition to being a purely technical attempt to fully quantify the relevant factors involved in the evaluation of health technologies in a transparent way, the gGEM is also a new way of thinking that allows healthcare budget holders to reward innovation in an economically viable way to promote the collective benefit.

In practice, the value of any such model must be determined in relation to the degree to which it understands the process of choosing, as well as to the degree to which it constitutes an improved analysis tool for practical applications. Such applications are also important beyond the field of genetics, especially for countries with truly limited available resources that need to be used prudently. We should mention that this model has been used as a tool of analysis for the economic evaluation of health services relevant to genetics, but it can be used for the healthcare system in general or in a broader economic sector as long as cost and effectiveness can be evaluated discretely.

In conclusion, the process of resource allocation is critical to allow a society to effectively exploit advancements in healthcare services. The evaluation of resource allocation strategies plays a major part in the maximization of social utility. This chapter presents the framework for resource allocation methods for genomic medicine as opposed to the classic model used by economists.

References

[1] Drummond M, Jonsson B, Rutten F. The role of economic evaluation in the pricing and reimbursement of medicines. Health Policy 1997;40(3):199–215.

[2] Fragoulakis V, Mitropoulou C, Williams MS, Patrinos GP. Economic evaluation in genomic medicine. Burlington, CA, USA: Elsevier/Academic Press; 2015. ISBN 978-0128014974.

[3] Weinstein MC, Torrance G, McGuire A. QALYs: the basics. Value Health 2009;12(Suppl 1):S5–9.

[4] Sullivan R, Peppercorn J, Sikora K, et al. Delivering affordable cancer care in high-income countries. Lancet Oncol 2011;12(10):933–80.

[5] O'Brien BJ, Briggs AH. Analysis of uncertainty in health care cost-effectiveness studies: an introduction to statistical issues and methods. Stat Methods Med Res 2002;11(6):455–68.
[6] Barton GR, Briggs AH, Fenwick EA. Optimal cost-effectiveness decisions: the role of the cost-effectiveness acceptability curve (CEAC), the cost-effectiveness acceptability frontier (CEAF), and the expected value of perfection information (EVPI). Value Health 2008;11(5):886–97.
[7] Donaldson C, Birch S, Gafni A. The distribution problem in economic evaluation: income and the valuation of costs and consequences of health care programmes. Health Econ 2002;11(1):55–70.
[8] Donaldson C, Currie G, Mitton C. Cost effectiveness analysis in health care: contraindications. Br Med J 2002;325(7369):891–4.
[9] Eckermann S, Briggs A, Willan AR. Health technology assessment in the cost-disutility plane. Med Decis Mak 2008;28(2):172–81.
[10] Gafni A. Willingness to pay. What's in a name? PharmacoEconomics 1998;14(5):465–70.
[11] Whitehead SJ, Ali S. Health outcomes in economic evaluation: the QALY and utilities. Br Med Bull 2010;96:5–21.
[12] Snyder SR, Mitropoulou C, Patrinos GP, Williams MS. Economic evaluation of pharmacogenomics: a value-based approach to pragmatic decision making in the face of complexity. Public Health Genom 2014;17(5-6):256–64.
[13] Mitropoulou C, Fragoulakis V, Bozina N, Vozikis A, Supe S, Bozina T, Poljakovic Z, van Schaik RH, Patrinos GP. Economic evaluation of pharmacogenomic-guided warfarin treatment for elderly Croatian atrial fibrillation patients with ischemic stroke. Pharmacogenomics 2015;16(2):137–48.
[14] Fragoulakis V, Mitropoulou C, van Schaik RH, Maniadakis N, Patrinos GP. An alternative methodological approach for cost-effectiveness analysis and decision making in genomic medicine. OMICS 2016;20 (5):274–82.
[15] O'Brien BJ, Gertsen K, Willan AR, Faulkner LA. Is there a kink in consumers' threshold value for cost-effectiveness in health care? Health Econ 2002;11(2):175–80.
[16] Vernon JA, Goldberg R, Golec J. Economic evaluation and cost-effectiveness thresholds: signals to firms and implications for R&D investment and innovation. PharmacoEconomics 2009;27(10):797–806.
[17] Fragoulakis V, Mitropoulou C, Katelidou D, van Schaik RH, Maniadakis N, Patrinos GP. Performance ratio based resource allocation decision-making in genomic medicine. OMICS 2017;21(2):67–73.
[18] Lubbe W. Should IQWiG revise its methods of cost-effectiveness analysis in order to comply with more widely accepted health economical evaluation standards? Dtsch Med Wochenschr 2010;135(12):582–5.
[19] Briggs A, Sculpher M, Buxton M. Uncertainty in the economic evaluation of health care technologies: the role of sensitivity analysis. Health Econ 1994;3(2):95–104.
[20] Flessa S. Where efficiency saves lives: a linear programme for the optimal allocation of health care resources in developing countries. Health Care Manag Sci 2000;3(3):249–67.

CHAPTER

8 Using "big data" for economic evaluations in genomics

Sarah Wordsworth[a,b], Brett Doble[c], Katherine Payne[d], James Buchanan[a,b], Deborah Marshall[e], Christopher McCabe[f], Kathryn Philips[g], Patrick Fahr[a,b], and Dean A. Regier[h,i]

[a]*Health Economics Research Centre, Nuffield Department of Population Health, University of Oxford, Oxford, United Kingdom,* [b]*Oxford National Institute for Health Research Biomedical Research Centre, Oxford, United Kingdom,* [c]*Paraexel, London, United Kingdom,* [d]*Manchester Centre for Health Economics, Division of Population Health, Health Services Research & Primary Care, School of Health Sciences, The University of Manchester, Manchester, United Kingdom,* [e]*Department of Community Health Sciences, University of Calgary, Calgary, AB, Canada,* [f]*Queen's Management School, Queen's University, Belfast, Northern Ireland,* [g]*Department of Clinical Pharmacy, Center for Translational & Policy Research on Personalized Medicine (TRANSPERS), University of California, San Francisco, CA, United States,* [h]*Cancer Control Research, BC Cancer, Vancouver, BC, Canada,* [i]*School of Population and Public Health, University of British Columbia, Vancouver, BC, Canada*

8.1 Introduction

In genomics, especially in diagnostics, less randomized controlled trials (RCTs) are undertaken compared to the evaluation of medicines. Health economists therefore have fewer opportunities to collect the resource use, cost, and outcome information needed to undertake economic evaluations of genomic technologies. One potential alternative data source are observational studies, which can create so-called big data. "Big data" is often described as having 5 "Vs": volume, velocity, variety, veracity, and value [1]. Big datasets containing information for millions of people on their health records, genome sequencing results, imaging, and other publicly available data present an opportunity to expand the digital element of healthcare [2]. Uses for big data in health care include prediction (diagnostic or prognostic) models and observational studies comparing healthcare interventions [3]. Although the use of big data for health care is still quite new, evidence suggests that such datasets already offer moderate-to-high accuracy for the diagnosis of some diseases and may improve the management of some chronic diseases [4]. Several areas of health care have already made use of big data to inform clinical decision-making, such as intensive care [5], accident and emergency [6], cardiovascular diseases [2], and oncology [7]. It has also been suggested that big data analytics could produce benefits for the economy [8].

Considerable funding has been invested in trying to reap the potential benefits of big data. In Europe, several funding calls and projects in the Horizon 2020 program have focused on the use of big data for better health care. For example, BigMedilytics (Big Data for Medical Analytics) is a large European Commission funded initiative aimed at trying to transform the region's healthcare sector

Economic Evaluation in Genomic and Precision Medicine. https://doi.org/10.1016/B978-0-12-813382-8.00008-2

by using state-of-the-art big data technologies to reduce cost, improve patient outcomes, and improve access to healthcare. The project has twelve wide-ranging pilot studies exploring population health, chronic disease management, oncology, and industrialization of healthcare services, covering disease prevention, diagnosis, treatment, and home care (www.bigmedilytics.eu/big-data-project). In the United States, the All of US Research Program is sequencing to detect a whole range of diseases (common and rare) to build a core data set by enrolling a diverse group of participants who are completing questionnaire responses and data are collected from electronic health records (EHRs), biospecimens and physical measurements for many years to enable the study of biological, social, and environmental determinants of health and disease in a precision medicine context (https://allofus.nih.gov). Real World Data (RWD) provided by observational studies, including medical claims data, EHRs, surveys and disease registries are being used more often to provide information for economic evaluations. Although observational studies have been considered as less robust [9] methodologically than RCTs, there is growing recognition that RWD from observational studies can be a viable alternative to RCTs and can be used to help extrapolate trial results beyond the end of the trial or to extrapolate to a larger population that the trial in the context of real word health care [10]. Real world evidence (RWE) is evidence on the benefits, risks, and cost-effectiveness of health technologies based on RWD [9]. Genomics is an area where RWD and RWE are likely to be a useful source of information, with precision oncology noted as a potential use for RWD in the context of health technology assessment specifically [11].

Genome sequencing is an important example of "big data" because it produces large amounts of complex data [12]. From a health economics perspective, genomics-based big datasets could provide valuable information to help study the cost-effectiveness of identifying rare disease variants in scenarios in which an RCT is unlikely to be feasible, for example, due to very small sample sizes [13], especially for rare genetic diseases. The big data generated in these large sequencing programmes will hopefully provide evidence of the benefits of sequencing for healthcare systems and could be used to populate cost-effectiveness models assessing alternative sequencing technologies (such as panels, whole exome sequencing, whole genome sequencing and possibly future technologies).

This chapter examines how "big data" could be useful in the economic evaluations of genomic tests. It highlights the main methodological and practical challenges that health economists and other researchers may encounter when using big data for economic evaluations and suggests potential solutions for these challenges. The chapter is based on a paper by Wordsworth et al. [14].

8.2 Using "big data" for the economic evaluation of genomic tests

The scope for using big data to inform cost-effectiveness analyses of sequencing technologies is expanding. Several large genome sequencing initiatives have extensive genomic and health economic data (especially cost data) that could be used to help determine the cost-effectiveness of genome sequencing relative to other types of genomic and non-genomic tests. In the United Kingdom, the 100,000 Genomes Project was the largest genomic sequencing initiative to routinely collect linked healthcare and sequencing data at scale for patients with rare diseases or cancer (www.genomicsengland.co.uk/about-genomics-england/the-100000-genomes-project/). This study, which was completed in December 2018, linked whole-genome sequencing (WGS) data to patient-level longitudinal data extracted from routine healthcare databases in the UK national health service (NHS), including secondary care records (Hospital Episode Statistics; HES), disease registries, pharmacy data, and mortality data. The

All of Us Research Program in the United States is sequencing one million individuals for a wide range of diseases, both common and rare, as well as healthy individuals (https://allofus.nih.gov). It is anticipated that this program will yield an electronic core dataset that will contain participant-provided information on sociodemographic variables, electronic health record data, and potentially mobile and digital health data. One of the key aims of this program is to help build one of the most diverse health databases of its kind (https://allofus.nih.gov/). A further major study, the UK Biobank, is releasing the whole genome sequencing (WGS) data for its 500,000 participants, representing the world's largest release of WGS data. When combined with other study data, such as lifestyle, biochemical, imaging and health outcome data, once linked with health care resource use data, it will enable researchers to gain an insight into the role of genomic data and provide information to examine resource use and cost implications for different diseases and health care interventions (https://www.ukbiobank.ac.uk/learn-more-about-uk-biobank/news/whole-genome-sequencing-data-on-200-000-uk-biobank-participants-available-now, https://www.ukbiobank.ac.uk/enable-your-research/about-our-data/genetic-data).

The big data that will be generated by these sequencing programs provides evidence of the benefits of sequencing for healthcare systems and could be used to populate cost-effectiveness models assessing alternative sequencing technologies (e.g., panels, whole exome sequencing, and whole genome sequencing).

8.3 Analytics for omics data

In terms of the actual use of big data in the context of economic evaluations, a scoping review by Bakker et al. [8] looked into the health and economic impact of big data analytics for clinical decision-making. The authors found that of 71 studies meeting their eligibility criteria, few were full economic evaluations where both cost and outcome data were collected and brought together. Twenty studies specifically examined "big data analytics," but only seven reported both cost savings and improved outcomes from using big data [8]. Eleven papers reported the potential impact of predictive and prescriptive analytics of omics data, often with the aim of applying them as a test in clinical practice. Only two papers focused specifically on the use of sequencing data and one paper combined multiple types of data (pharmacogenomics, literature, medical history). The remaining papers examined microarray data, and all the analytics that were adopted as a test were used in oncology [8]. The authors concluded that "The promised potential of big data is not yet reflected in the literature, partly because only a handfull of apropriately performed economic evaluations have been published." They also noted that there was no clear definition of "big data," which prevented policy makers and healthcare professionals from determining which big data initiatives are worth implementing [8].

8.4 Difficulties in using "big data" in for economic evaluation of sequencing tests

Using big data for the economic evaluation of sequencing technologies has numerous challenges. There are three areas where challenges have been noted: data collection, data management, and data analysis [14]. We describe these challenges and suggests some possible solutions in the sections that follow.

8.4.1 Data collection challenges

8.4.1.1 Small number of observations

Large scale sequencing initiatives provide an opportunity to maximize the "volume" of data that can be obtained for specific conditions. This may be helpful for common diseases, but the incidence and prevalence of rare diseases is much lower and patterns in disease progression and treatment response are difficult to identify, even when recruiting from population-based studies. Large scale sequencing initiatives provide an opportunity to maximize the 'volume' of data that can be obtained for specific conditions. This may be helpful for specific cancer types, but the incidence and prevalence of rare diseases is much lower and patterns in disease progression and treatment response are difficult to identify, even when recruiting from population-based studies. To help deal with small sample sizes, N-of-1 trials can be an option. N-of-1 studies are randomized, controlled, multiple crossover trials which examine treatment effects in single patients, unlike RCTs where average treatment effects are examined. They enable treatments to be modified (personalised) to the individual and if data can be aggregated from numerous N-of-1 trials, this can increase the amount of data available for economic evaluations [15].

However, in practice, very few N-of-1 trials have been carried out. Techniques to aggregate outcomes from multiple N-of-1 trials in a way that is not subject to allocation or confounding bias are required if such data are to inform economic evaluations [16]. An alternative approach for dealing with small sample sizes is to combine data from sequencing initiatives conducted in multiple countries. Although aggregating data across different health systems could increase the risk of unknown confounding, and present cross-border ethical and data access challenges, this is an option that major sequencing companies are trying to harness (www.illumina.com).

8.4.1.2 Insufficient outcomes data for cost-effectiveness analysis

Information on health outcomes such as health related quality of life is crucial for evaluating the cost-effectiveness of genomic technologies. Unfortunately, quality of life data are rarely collected routinely in administrative healthcare datasets, and few of the recently completed and ongoing major sequencing initiatives are collecting such information. The Australian Cancer 2015 Pilot Study was one of the few big data sequencing studies to have collected this data [17]. More commonly, diagnostic yield has been used as an outcome measure in economic evaluations using big data to evaluate sequencing, as this outcome measure is more frequently available in large sequencing studies.

Published data sources for utility data can be used if primary data on quality of life are not available. The Cost-Effectiveness Analysis (CEA) Registry is a comprehensive database of over 10,000 cost-utility analyses on a wide variety of diseases and treatments published from 1976 onwards (https://cevr.tuftsmedicalcenter.org/databases/cea-registry). However, there is limited utility data for rare diseases in the database. In the future, in clinical areas where limited or no published utility data are available, it would be helpful if appropriate health-related quality of life questionnaires and other outcome measurement tools could be completed by participants in genomic-linked cohort studies.

8.4.2 Data management challenges

8.4.2.1 Data linkage

In a sequencing context, big data is usually obtained by linking sequencing data to multiple data sources from several settings (e.g., secondary care, primary care, disease registries, and mortality datasets). These data are often collected by different stakeholders and accessed via different platforms.

Stakeholders commonly have turnaround times of more than a month to release data because data linkage is performed in house (due to privacy and legislative requirements). Therefore, the linked datasets received by researchers may be outdated on receipt. Linking multiple data sources from different stakeholders can also understate real-world variability, as only a certain proportion of patients may have all the desired data elements [14]. For example, in the analysis of pilot data from Cancer 2015 only a proportion (621/922; 67%) of the total sample with resource use data also had sequencing data. This limited the analyses to these data subsets, potentially biasing any economic evaluations that use these data [17].

Difficulties encountered with having multiple stakeholders in data linkage processes could potentially be mitigated by using a central data warehouse of secure servers. However, this requires the coordination of multiple stakeholders, some of whom may not have a research mandate. Even if such systems were in place, many routinely collected databases have a time lag between the occurrence of a healthcare event and the availability of this data in the dataset (e.g., for HES data in the United Kingdom there is approximately a six-month delay). Indeed, in the 100,000 Genomes Project, primary care data were not available in the secure research platform for the project until several years after the main project had completed.

Current best practice where data are missing for a subset of patients is to use imputation methods (such as multiple imputation) that can operate with partial data to avoid dropping observations and introducing bias [18]. Such approaches could help to resolve issues related to missing data due to linkage problems. However, multiple imputation may not be appropriate if entire datasets or data elements are missing. Furthermore, some imputation approaches (e.g., "predictive mean matching"), may not work well when patient subgroups are very small, for example, with rare genetic diseases.

In terms of privacy concerns, limiting data access for researchers to certain variables by using a pre-approved protocol could help to minimize re-identification risks [17] and could be framed as a minimum or core dataset to answer a specific research question [19]. Such approaches are commonly used to access routinely collected administrative healthcare datasets. This is particularly important as, even with consent, patients may not be totally aware what their data are being used for.

8.4.2.2 Large number of zero observations

A common problem with the resource use aspect of economic evaluation of big data is a high level of zero observations. This is where there is no resource use recorded for many people and can create problems for analysis when trying to examine variations in resource use and costs, especially if the population being studied is relatively healthy. For example, healthcare resource use prior to a cancer diagnosis is likely to be minimal given that most individuals are relatively healthy at that point and may only begin to consume healthcare resources after receiving a cancer diagnosis [17].

One possible solution is to collapse the data into longer periods to reduce the number of zero observations. However, this can create problems if important data patterns are hidden. If there are high levels of zero observations (>50%), alternative econometric models that account for zeros, such as zero-inflated and hurdle models could be used.

8.4.2.3 Difficulties in setting up big datasets for economic evaluation

The datasets linked to create big data are generally not designed for research (although the UK 100,000 Genomes Project and the US All of US Program are exceptions), and they might not provide the data in a way that is straightforward to use in an economic evaluation. For example, diagnostic information may be lacking due to differences in how elements are described and subsequently coded. Also, routinely collected administrative healthcare datasets are often incomplete, inconsistent, and contain errors. To produce big data that has acceptable veracity, these data must be cleaned. However, data cleaning requires numerous steps that may further reduce the veracity of the final dataset. For example, assigning unit costs to HES data requires analysts to remove duplicates, organize the data into unique finished consultant episodes, and then adjust the variables to enable linkage to Healthcare Resource Group (HRG) codes and National Reference Costs. Furthermore, the coding of such datasets often changes over time [14].

One way to enhance big datasets to better inform economic analyses is to develop a set of clinical rules based on the review of carefully selected standard cases for different phenotypes of interest. A first step in such a review could be "visualization" of the patient pathways using schematics detailing the main events for individual patients over the entire period of available data. Algorithms developed using this approach would need to be thoroughly validated to ensure veracity. To avoid issues related to coding changes over time, it may be necessary to standardize data to a common year to facilitate linkage to unit costs. This would help to ensure that cost variations across years reflect changes in utilization rather than coding changes.

8.4.3 Data analysis challenges

8.4.3.1 Selection bias and confounding

Using representative databases when selecting datasets for linkage can cause selection bias and confounding. For example, in the 100,000 Genomes Project, hospital episide statistics data were only available for England, not the entire United Kingdom. Using more data of the same type to improve prediction does not address issues related to bias from missing variables or selection bias in the data. Although this issue is not specific to the evaluation of sequencing, it becomes more acute when observational databases are being used as a key data source for estimating causal relationships in economic evaluations of genomic sequencing.

One potential solution is to use an instrumental variables approach, where the relationship between an exposure (e.g., a specific disease) and an outcome (e.g., healthcare costs) can be confounded by either observed or unobserved confounders. Consequently, getting an unbiased estimate of the causal effect of an exposure on an outcome using observational data can be problematic. However, the increased availability of genomic data linked to longitudinal data on healthcare resource use and costs provides an opportunity to better understand such causal relationships. Since the allocation of genomic variants within a population is random for many diseases, the effects of an endogenous variable (disease) on an outcome (healthcare costs) can be identified by using an instrumental variable analysis [20]. Genomic variants are potentially useful instruments as they can be associated with a disease but not necessarily with potential confounders, and it is possible that identified genomic variants will not directly impact costs other than via the disease in question [20].

Alternatively, propensity score matching could be used to ensure the design of an observational study is analogous to that of a randomized experiment. This approach is effective because researchers do not see information on outcome variables until after a study has finished. Because propensity scores are a function of covariates, not outcomes, repeated analyses that try to balance covariate distributions across treatment groups should not bias estimates of the treatment effect on outcome variables [21].

8.4.3.2 Lack of a counterfactual

Economic evaluations require a comparator intervention/s. However, it is not always straightforward to identify a counterfactual for sequencing technologies, especially in a big data context. One potential solution is to identify a matching cohort to differentiate costs associated with a particular genetics disease from costs associated with routine healthcare services. For example, in the context of rare diseases in children in Canada, the direct healthcare costs associated with the care of children with selected genetic diseases was compared to three matched cohorts: two cohorts of children with chronic disease (asthma and diabetes) and one cohort of children from the general population. The index event date for patients with a genetic disease was defined as the date of diagnosis (determined from a chart review). To compare cohorts, the comparison cohorts were matched to the genetic disease cohort based on sex, date of birth, income quintile, rural versus urban household at birth, and index event date [22].

"Umbrella" and "basket" trials are not randomized, but could be potential alternatives to RCTs [23]. Umbrella trials enroll patients with a single tumor type or histology who are then treated according to the molecular characterization of each case. These trials involve sub-studies that are connected through a central infrastructure overseeing patient identification and screening. This approach is particularly beneficial for low-prevalence markers and allows the testing of new drugs and biomarkers. Basket trials also have a central screening and treatment infrastructure, but in contrast to umbrella trials, these trials facilitate the study of multiple molecular subpopulations of different tumor or histologic types within a single study and can include highly rare cancers that would be difficult to study in RCTs [23]. The basket design has the flexibility to open and close study arms; hence, several drugs for many different diseases can be screened.

8.4.4 Identifying key challenges

As there are multiple challenges associated with the use of big data in genomics, it is helpful to understand how frequent the challenges are encountered in practice. A review by Fahr et al. [24] attempted to identify and summarize the main challenges reported in the literature. The review included papers that examined issues relevant to the interconnectedness of biomedical big data, precision medicine, and health economic evaluation. The challenges identified related to data management, data quality, and data analysis. The review concluded reassuringly that most of the challenges such as gaining access to large volumes of data from multiple sources and trying to link the data in the context of complcated data access and sharing procedures, are essentially practical issues that could be overcome with improved procedures for data access and sharing. However, the authors noted that the existence of missing data across linked datasets, and data quality issues in particular might need an evolution in economic evaluation methods, which is less encouraging [24].

8.5 Conclusions

The use of big genomic datasets to undertake economic evaluations provides hope for the future. However, health economists and other researchers are likely to encounter challenges when using these data. Some of these challenges are common to most cost-effectiveness analyses using observational big data, but others are unique or more pronounced with genomic sequencing studies. In the future, health economists who work with big data in the context of genome sequencing should clearly document the key challenges they face and report the successes (or otherwise) from the solutions they have applied to their challenges [24,25].

References

[1] Mehta N, Pandit A. Concurrence of big data analytics and healthcare: a systematic review. Int J Med Inform 2018;114:57–65.

[2] Hemingway H, Asselbergs FW, Danesh J, et al. Big data from electronic health records for early and late translational cardiovascular research: challenges and potential. Eur Heart J 2018;39(16):1481–95.

[3] Collins B. Big data and health economics: strengths, weaknesses, opportunities and threats. PharmacoEconomics 2016;34:101–6.

[4] Borges do Nascimento IJ, Marcolino MS, Abdulazeem HM, Weerasekara I, Azzopardi-Muscat N, Gonçalves MA, et al. Impact of big data analytics on people's health: overview of systematic reviews and recommendations for future studies. J Med Internet Res 2021;23(4), e27275. https://doi.org/10.2196/27275.

[5] Sanchez-Pinto LN, Luo Y, Churpek MM. Big data and data science in critical care. Chest 2018;154 (5):1239–48.

[6] Janke AT, Overbeek DL, Kocher KE, et al. Exploring the potential of predictive analytics and Big Data in emergency care. Ann Emerg Med 2016;67(2):227–36.

[7] Fronhlich H, Balling R, Beerenwinkel N, et al. From hype to reality: data science enabling personalized medicine. BMC Med 2018;16(1):150.

[8] Bakker L, Aarts J, Uyl-de Groot C, Redekop W. Economic evaluations of big data analytics for clinical decision-making: a scoping review. J Am Med Inform Assoc 2020;27(9):1466–75. https://doi.org/10.1093/jamia/ocaa102. PMID: 32642750; PMCID: PMC7526472.

[9] Bowrin K, Briere JB, Levy P, Millier A, Clay E, Toumi M. Cost-effectiveness analyses using real-world data: an overview of the literature. J Med Econ 2019 Jun;22(6):545–53. https://doi.org/10.1080/13696998.2019.1588737.

[10] Pietri G, Masoura P. Market access and reimbursement: the increasing role of real-world evidence. Value Health 2014;17:A450–1.

[11] Regier DA, Pollard S, McPhail M, et al. A perspective on life-cycle health technology assessment and real-world evidence for precision oncology in Canada. npj Precis Onc 2022;6:76. https://doi.org/10.1038/s41698-022-00316-1.

[12] Phillips KA, Trosman JR, Kelley RK, et al. genomic sequencing: assessing the health care system, policy, and big-data implications. Health Aff 2014;33:1246–53.

[13] Chen Y, Guzauskas GF, Gu C, et al. Precision health economics and outcomes research to support precision medicine: big data meets patient heterogeneity on the road to value. J Personal Med 2016;6.

[14] Wordsworth S, Doble B, Payne K, Buchanan J, Marshall DA, McCabe C, Regier DA. Using "Big Data" in the cost-effectiveness analysis of next-generation sequencing technologies: challenges and potential solutions. Value Health 2018;21(9):1048–53. https://doi.org/10.1016/j.jval.2018.06.016. Epub 2018 Aug 17.PMID: 30224108.

[15] Grammatikopoulou MG, Gkouskou KK, Gkiouras K, et al. The niche of *n*-of-1 trials in precision medicine for weight loss and obesity treatment: back to the future. Curr Nutr Rep 2022;11:133–45. https://doi.org/10.1007/s13668-022-00404-5.

[16] Doble B, Harris A, Thomas DM, et al. Multiomics medicine in oncology: assessing effectiveness, cost-effectiveness and future research priorities for the molecularly unique individual. Pharmacogenomics 2013;14:1405–17.

[17] Lorgelly PK, Doble B, Knott RJ, et al. Realising the value of linked data to health economic analyses of cancer care: a case study of cancer 2015. PharmacoEconomics 2016;34:139–54.

[18] White IR, Royston P, Wood AM. Multiple imputation using chained equations: issues and guidance for practice. Stat Med 2011;30:377–99.
[19] Pollard S, Weymann D, Chan B, Ehman M, Wordsworth S, Buchanan J, et al. Defining a core data set for the economic evaluation of precision oncology. Value Health 2022;25(8):1371–80. https://doi.org/10.1016/j.jval.2022.01.005.
[20] Dixon P, Hollingworth W, Harrison S, Davies NM, Davey SG. Mendelian randomization analysis of the causal effect of adiposity on hospital costs. J Health Econ 2020;70, 102300. https://doi.org/10.1016/j.jhealeco.2020.102300. Epub 2020 Jan 25. PMID: 32014825; PMCID: PMC7188219.
[21] Rubin DB. Using propensity scores to help design observational studies: application to the tobacco litigation. Health Serv Outcome Res Methodol 2001;2:169–88.
[22] Marshall DA, Benchimol EI, MacKenzie A, et al. Direct health-care costs for children diagnosed with genetic diseases are significantly higher than for children with other chronic diseases. Genet Med 2019;21(5):1049–57. https://doi.org/10.1038/s41436-018-0289-9. PMID: 30245512.
[23] Rashdan S, Gerber DE. Going into BATTLE: umbrella and basket clinical trials to accelerate the study of biomarker-based therapies. Ann Transl Med 2016;4.
[24] Fahr P, Buchanan J, Wordsworth S. A review of the challenges of using biomedical big data for economic evaluations of precision medicine. Appl Health Econ Health Policy 2019;17(4):443–52. https://doi.org/10.1007/s40258-019-00474-7. PMID: 30941659; PMCID: PMC6647451.
[25] Hayeems RZ, Bernier F, Boycott KM, Hartley T, Michaels-Igbokwe C, Marshall DA. Positioning whole exome sequencing in the diagnostic pathway for rare disease to optimise utility: a protocol for an observational cohort study and an economic evaluation. BMJ Open 2022;12(10), e061468. https://doi.org/10.1136/bmjopen-2022-061468.

Further reading

Craig P, Cooper C, Gunnell D, et al. Using natural experiments to evaluate population health interventions: new MRC guidance. J Epidemiol Community Health 2012;66:1182–6.
European Commission (EC). EU countires will cooperate in linking genomic databases across borders. Brussels: EC; 2018.
Genomics England. The 100,000 genomes project. London: Department of Health & Social Care; 2018.
National Institutes of Health. About the all of us research program. Washington: Department of Health and Human Services; 2018.
Schwarze K, Buchanan J, Taylor JC, et al. Are whole-exome and whole-genome sequencing approaches cost-effective? A systematic review of the literature. Genet Med 2018;20.
Tufts Medical Centre. Cost-effectiveness analysis registry. Boston: Tufts Medical Center; 2019.

CHAPTER

9 Assessing the stakeholder environment and views towards implementation of personalized medicine in a healthcare setting

Christina Mitropoulou[a,b], Athanassios Vozikis[c], and George P. Patrinos[b,d,e]

[a]*The Golden Helix Foundation, London, United Kingdom,* [b]*Department of Genetics and Genomics, College of Medicine and Health Sciences, United Arab Emirates University, Al-Ain, United Arab Emirates,* [c]*Economics Department, University of Piraeus, Piraeus, Greece,* [d]*Department of Pharmacy, School of Health Sciences, University of Patras, Patras, Greece,* [e]*Zayed Center of Health Sciences, United Arab Emirates University, Al-Ain, United Arab Emirates*

9.1 Introduction

Personalized medicine could help to optimize medical decision-making by informing patient-specific treatment decisions to the benefit of both patients and national healthcare systems. This is achieved by combining information on an individual's unique genomic profile with other data related to proteomics or metabolomics, for example. Such information enables healthcare professionals to make tailor-made disease and treatment risk assessments [1], individualizing therapeutic interventions [2].

The personalized medicine environment is complex; multiple stakeholders with varying levels of genetics knowledge intersect with each other. Previous studies have described genetics awareness in the general public, assessed the genetics education level of healthcare professionals [3], and described the views of these professionals on ethical, legal, and societal (ELSI) issues pertaining to (pharmaco) genomics and personalized medicine [4–8]. These studies, conducted in both Europe and the United States, have shown that public awareness of genomic and personalized medicine is often low. The latter can be a major barrier to expediting the implementation of personalized medicine, since the public is often skeptical to embrace an innovative approach, particularly with regards to health care [9]. To better understand and analyze the genomic and personalized medicine policy environment in a given country and healthcare system, a stepwise approach is required that comprehensively identifies all key stakeholders, summarizes their interests and motivations, and describes the main opportunities and obstacles for policy implementation.

This chapter applies such a stepwise approach in the fields of genomic and personalized medicine. Specifically, we identify the key stakeholders involved in the implementation of genomic and personalized medicine, describe how their views, interests, and opinions can be elicited, and consider their potential involvement in taking personalized medicine from discovery to the clinic. We also outline

Economic Evaluation in Genomic and Precision Medicine. https://doi.org/10.1016/B978-0-12-813382-8.00002-1

previously conducted work to assess such a personalized medicine environment, with a view to understanding the challenges related to personalized medicine and identifying ethical, legal, and regulatory deficiencies that must be rectified.

9.2 Identifying stakeholders in personalized medicine

The key stakeholders in the field of personalized medicine can be identified mainly via expert consultation involving experts from the academia, industry, and governmental organizations, as well as thorough literature search and analyses, such as systematic reviews and meta-analyses. The following key stakeholders exist in the personalized medicine policymaking environment:

(a) academic and research organizations, such as public and private universities, colleges, and research institutes,
(b) public and private genetic laboratories, including those affiliated with hospital clinics,
(c) physicians who specialize in genetics, namely, clinical and laboratory geneticists,
(d) physicians from other medical specialties,
(e) payers, including both public health insurance funds and private health insurance companies,
(f) genetics and genomics professional associations, including clinical and laboratory geneticists and genetic counseling professionals,
(g) pharmaceutical and biotechnology companies, especially those with previous involvement in personalized medicine,
(h) national medicines organizations that regulate pharmacogenomic and personalized medicine services,
(i) national ministries of health,
(j) national and/or regional bioethics councils,
(k) religious organizations,
(l) other companies, such as those who sell reagents, service providers, and manufacturers,
(m) pharmacies,
(n) consumers and citizens, and
(o) press and the media.

There may be additional stakeholders that could be considered when performing such analyses, in the context of different healthcare settings, whose views and opinions related to personalized medicine must be considered, using approaches as those outlined in the sections that follow.

9.3 Eliciting and analyzing stakeholder views and opinions related to personalized medicine

A key first step in the process of eliciting stakeholder views and opinions in this context is to collect the required data, mainly by either conducting structured interviews and/or distributing questionnaires, while the literature also involves another important data source. This data is subsequently analyzed using specific tools. According to our previous experience, it seems that the stakeholder mapping approach involving the PolicyMaker tool is a very convenient way for collecting and organizing important policy information and stakeholder views and opinions for genomic and personalized medicine. To our knowledge, there is no other approach that would categorize stakeholders and their views in such structured way to allow policy making, not only in personalized medicine but also in other healthcare-related disciplines.

PolicyMaker is a political mapping tool that serves as a database for assessments of policy content, the major stakeholders, their power of intervention, interests, and policy positions, and the networks and coalitions that interconnect them [10,11]. This tool aims to help policymakers to manage the processes of reform and promote strategic programming and thinking [12], potentially improving the political feasibility of proposed policy initiatives.

Use of the PolicyMaker tool requires analysts to follow five key design features of the tools, as also indicated at www.polimap.com/poliwhat.html:

(1) *Policy content*: To define and analyze the policy content, identify the major goals of the policy, specify a mechanism that is intended to achieve each goal, and determine whether the goal is already on the agenda.

(2) *Players/stakeholders*: To (a) identify the key players and most important stakeholders, (b) analyze their positions, views, power of intervention, and interests, (c) assess the policy's consequences for them, and (d) analyze the networks, coalitions and possibly synergies among the players.

(3) *Opportunities and obstacles*: To assess the opportunities and obstacles that affect the feasibility of a given policy by analyzing conditions within specific organizations and in the broader political environment.

(4) *Strategies*: To design strategies to improve the feasibility of a given policy measure by using expert advice provided in the program. Subsequently, it evaluates these strategies and creates alternative strategy packages as potential action plans.

(5) *Impacts of strategies*: To estimate the impacts of any given strategies on the positions, power, and number of mobilized stakeholders, which constitute the three main elements that affect the feasibility of a policy.

The analysis outcomes in PolicyMaker are presented in a series of tables and diagrams that systematically organize essential information about the policy in question. These tables and diagrams can be used in strategic planning for policy formulation and implementation to assist in improving the political feasibility of this policy (Fig. 9.1). The results can help with:

(a) analysis of the political circumstances faced by a given policy,

(b) rapid identification of bottlenecks and definition of obstacles,

(c) policymaking process, by assisting in communication among different organizations,

(d) data organization for storing, tracking, and analyzing positions, power, and other aspects of a political question,

(e) implementation strategies, by proposing new ideas and strategies, helping policymakers to evaluate their consequences, and to track down their implementation, and

(f) assessment of the overall impact and consequences of a given policy.

The output graphs and tables can be stored and compared to comparatively analyze current against previous positions, as outlined in the Position and Feasibility Graphs to show the impacts of an undertaken strategy (Fig. 9.2).

A

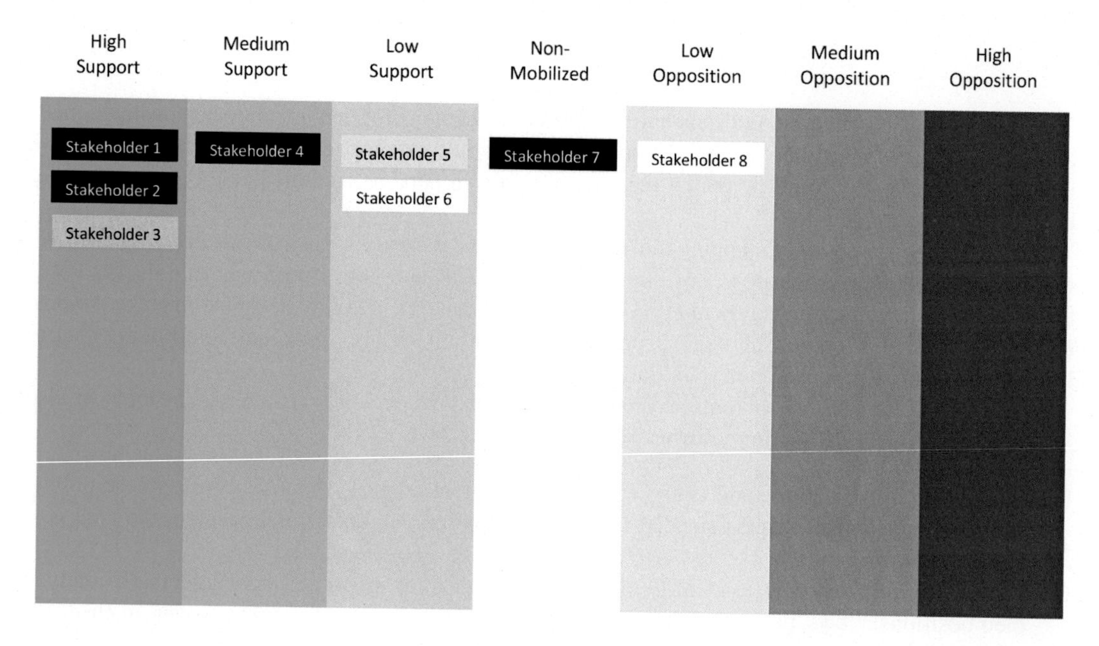

B

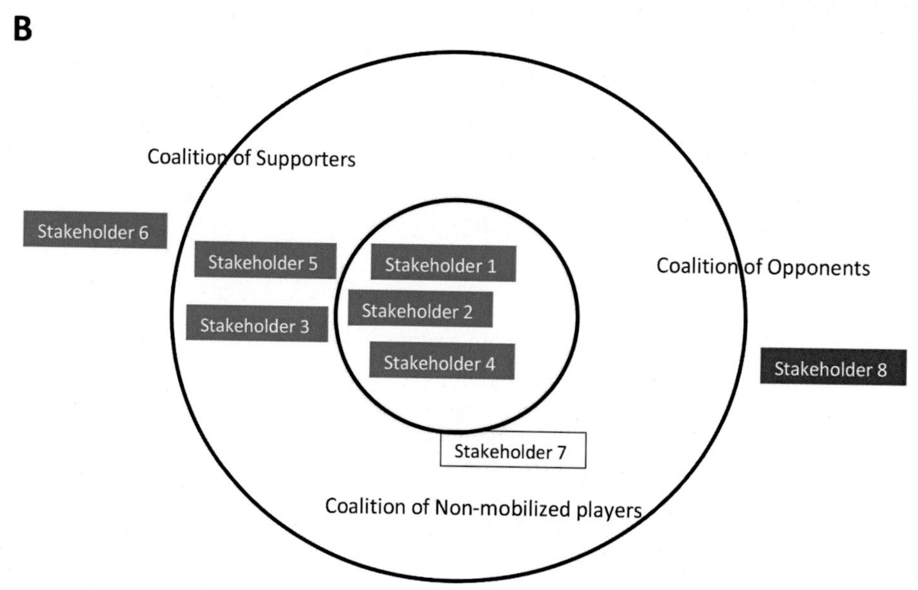

FIG. 9.1

Schematic drawing of the PolicyMaker analysis output. (A) The Current Position Map, in which each Stakeholder is grouped as Supporter (High, Medium, and Low Support, in different shades of *green*), Opponent (High, Medium, and Low Opposition, in different shades of *red*) and Non-Mobilized (in *white*). The power of intervention of each Stakeholder is depicted in *black* (High), *grey* (Medium), and *white* (Low). (B) The Coalition Map, depicting the stakeholders' initial position, various interests, and clustering, namely, coalition of supporters (*blue*; in the left), coalition of non-mobilized players (*white*; in the center), and coalition of opponents (*red*; in the right). Stakeholders are positioned in the same-centered circles according to their power of intervention (the higher the power of intervention, the more centrally the Stakeholder is placed). The Coalition Map is always evaluated jointly with the Current Position Map.

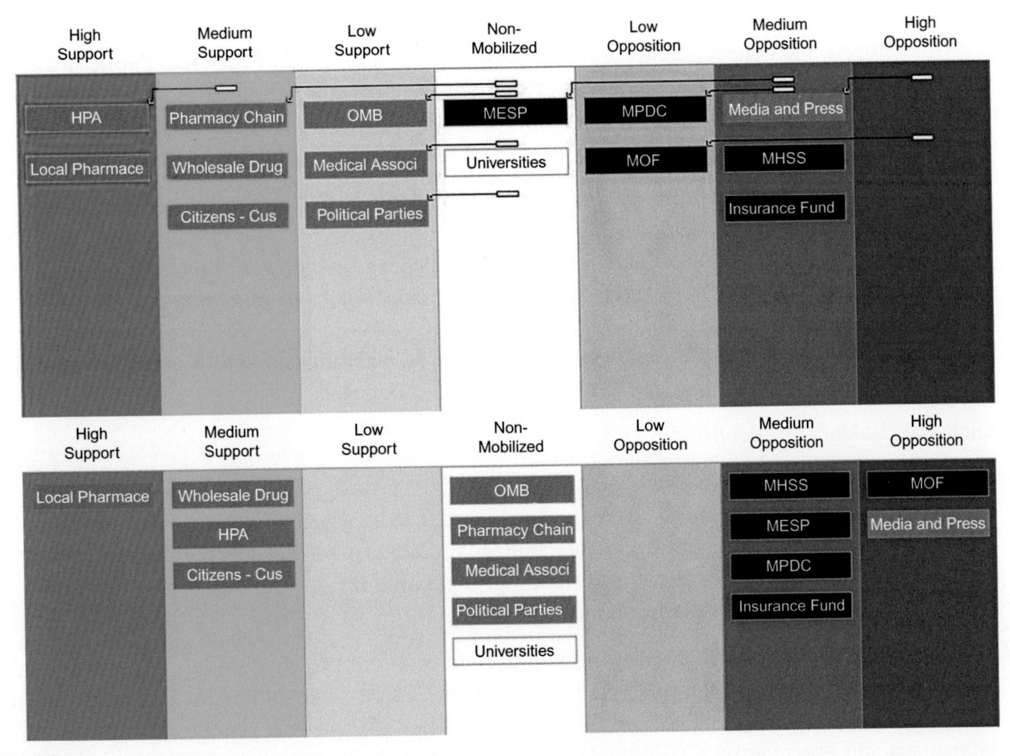

FIG. 9.2

Analysis of the Current Position Map against the previous positions of the same stakeholders (lower part of the figure). Note that the "OMB," "MedicalAssoci," and "PoliticalParties" stakeholders have been relocated from the "Non-Mobilized" to the "Low Support" group, the "MESP" stakeholder has been relocated from the "Medium Opposition" to the "Non-Mobilized" group, the "MPDC" stakeholder has been relocated from the "Medium Opposition" to the "Low Opposition" group, the "MOF" stakeholder has been relocated from the "High Opposition" to the "Low Opposition" group, and the "Media and Press" stakeholder has been relocated from the "High Opposition" to the "Medium Opposition" group. The latter findings depict a shift of various stakeholders towards a more supportive stance to the new policy under evaluation.

9.4 Implementing stakeholder analysis in genomic and personalized medicine: An example from the preliminary assessment of the genomic and personalized medicine environment in Greece

As mentioned in Section 9.1, limited information is available around the world on the views of key stakeholders regarding the implementation of genomic and personalized medicine. In most cases, this information is not systematically recorded, which makes it difficult to understand how these views have evolved over time. In this section we use the PolicyMaker tool to map the views of key stakeholders regarding the genomic and personalized medicine environment in Greece. We opted to undertake this study to map the current situation in Greece since the genomic and personalized medicine environment is poorly developed, the genetic testing services only scarcely provided in very few public hospitals, while the genetic testing services currently provided are not properly regulated by the existing legal framework, especially as far as direct-to-consumer genetic testing services are concerned.

Our approach has three steps:

(1) policy content analysis, including the goals and mechanisms for implementing genetic testing and genomic and personalized medicine,
(2) identification of the key stakeholders, analysis of their positions, power, interest and interrelationships, and assessment of the feasibility of current policies to implement genomic and personalized medicine, and
(3) determination of the strengths, weaknesses, opportunities, and threats that impact on the feasibility of these policies.

For each stakeholder, we identified:

(1) its current territorial level (national or regional)
(2) its sector (governmental, non-governmental, political, media, commercial, private, social)
(3) its position regarding the genomic and personalized medicine environment (high support, medium support, non-mobilized, medium opposition, high opposition)
(4) power of intervention (low, medium, high)

The views of some groups of stakeholders were identified from our previous studies and through their publicly expressed opinions in the media and conferences. These stakeholders included citizens and consumers (including 1717 members of the general public) [5], private genetic laboratories [13], and physicians and pharmacists (704 physicians of various medical specialties, 87 healthcare professionals (other than physicians), and 86 pharmacists) [5,6].

Based on these published findings and stated views during interviews and questionnaires, each stakeholder was clustered by the PolicyMaker tool according to (1) the extent of their support or opposition to genomic and personalized medicine in Greece (Fig. 9.3A) and (2) their position, including their various interests and their clustering (Fig. 9.3B).

Fifty percent of the key stakeholders were determined to be highly supportive of genomic and personalized medicine in Greece. Supportive stakeholders included pharmaceutical and biotechnology companies as well as molecular diagnostics laboratories. These stakeholders strongly support clinical implementation of pharmacogenomics from a technology-driven perspective.

Citizens, geneticists, other physicians, and pharmacies were also determined to be highly supportive of genomic and personalized medicine in Greece. These findings align with previous work that indicated the general population is in general positive towards individualized drug treatment modalities, even though their level of genetics awareness is fairly low [5,6,9,14].

Interestingly, both the Greek Ministry of Health and the public health insurance funds were determined to be in the "medium opposition" category. There are two main reasons for the medium opposition of these key stakeholders, both with very high power of intervention. First, the cost effectiveness of genome-guided treatment approaches is not yet fully proven in several European healthcare environments, including Greece among other countries. To overcome this obstacle, robust evidence on cost-effectiveness is required, generated via prospective clinical studies such as the Ubiquitous Pharmacogenomics project (www.upgx.eu) [15]. Second,

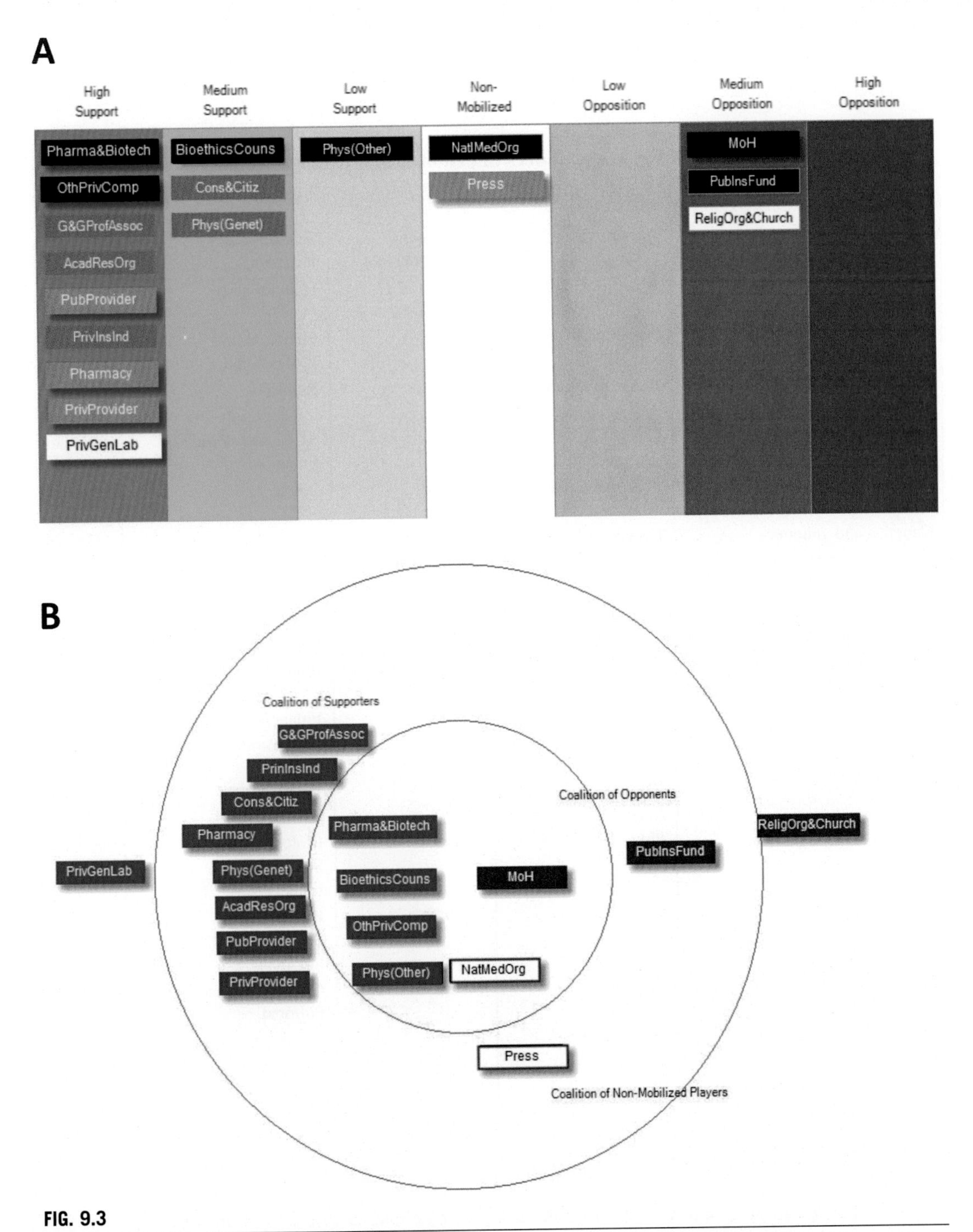

FIG. 9.3

(A) Current Position Map, depicting the stakeholders' current position on pharmacogenomics and genomic medicine. Black boxes depict "High Power" of the stakeholders to intervene, grey boxes "Medium Power," and white boxes "Low Power." (B) Coalition Map, providing a more comprehensive graphical presentation of the key stakeholders' current position, but also of the homogeneity of their interests and their grouping. For abbreviations, please refer to Table 9.1 (reproduced with permission).

Table 9.1 Characteristics of key stakeholders in genomic and personalized medicine in Greece.

Stakeholder	Abbreviations	Sector	Interest	Power
Academic and research organizations	AcadResOrg	Local non-governmental	Scientific and Financial	Medium
Greek National Bioethics Council	BioethicsCouns	Local non-governmental	Humanitarian	High
Private genetic laboratories	PrivGenLab	Private	Financial and Professional	Low
Religious organizations and church	ReligOrg&Church	Religious	Religious	Low
Consumers and citizens	Cons&Citiz	Social	Self-interest and Financial	Medium
Pharmaceutical and biotechnology companies	Pharma&Biotech	Private	Financial	High
Genetics and genomics professional associations	G&GProfAssoc	Local non-governmental	Professional and Scientific	Medium
Ministry of Health	MoH	Governmental	Financial	High
Payers (Private Health Insurance Industry)	PrivInsInd	Private	Financial	Medium
Payers (Public Health Insurance Funds)	PubInsFund	Governmental	Financial	High
Other private companies[a]	OthPrivComp	Private	Financial, Professional, and Scientific	High
Pharmacies	Pharmacy	Private	Financial and Professional	Medium
Physicians (Geneticists)	Phys (Genet)	Private	Professional, Scientific, and Financial	Medium
Physicians (Others)	Phys (Others)	Private	Professional, Scientific, and Financial	High
Press and Media	(Press)	Media	Ideological and Political	Medium
Private providers	PrivProvider	Private	Financial	Medium
Public providers	PubProvider	Governmental	Financial	Medium
Greek National Medicines Organization	NatlMedOrg	Governmental	Scientific	High

[a] *In the field of genomic medicine and/or pharmacogenomics, including pharmaceutical and biotechnology companies.*

delivery of genome-guided treatment modalities requires capacity building, both in terms of infrastructure and human capital, and both stakeholders have a strong financial interest in this context (see Table 9.1). A key challenge is that more than 75% of physicians in Greece think the costs of genetic testing services should be reimbursed by insurance companies [6], but public health insurance funds lack evidence on how genetic testing could reduce healthcare expenditure by reducing adverse drug reactions. Furthermore, the scope for capacity building, namely, infrastructure for genetic analysis, enhancing genomics awareness and knowledge, and so on, has been limited by the Greek financial crisis during which GDP contracted by almost 20% in 4 years (2010–2013) and the unemployment rate increased to almost 24% [16]. Simultaneously, the healthcare system has struggled to rationalize licensing, pricing, and reimbursement systems for healthcare services, medicines, and medical devices. Surprisingly, private health insurance companies were determined to be highly supportive of genomic and personalized medicine in Greece based on their responses to our interviews. This finding requires further investigation and possibly exploitation to convince public insurance funds to also adopt a supportive attitude towards this emerging trend of genomic medicine.

The current lack of proper legislation to oversee the operation of private genetic testing laboratories [14,17] could also explain the medium opposition of these stakeholders. This, however, contradicts the position of the National Medicines Organization, which is currently non-mobilized. Lastly, religious organizations display medium opposition, although their power to intervene is lower than that of other stakeholders.

The media and the press currently hold a neutral position on genomic medicine. If this were to change to medium-to-high support, with these stakeholders presenting objective opinions and facts by academics, qualified professionals, and regulatory bodies, this could facilitate and expedite adoption of genomic and personalized medicine in Greece and potentially alter the position of governmental organizations that currently hold an opposition stance towards genomic medicine.

9.5 Defining opportunities and threats when implementing genomic and personalized medicine in Greece

Our analysis of stakeholder views also identified several opportunities and threats in the genomic and personalized medicine policymaking environment in Greece.

A key opportunity that arises from the implementation of genomic medicine is the ability to treat diseases in a personalized manner. Stakeholders noted that personalized medicine could both enable genome-guided treatment rationalization, minimizing adverse drug reactions and improving quality of life, and contribute to better disease prevention, allowing patients to optimize their health plans and decisions. Drug treatment individualization could also help to reduce the healthcare expenditure at the national level (see also Ref. [18]); lowering the incidence of adverse drug reactions reduces both the duration and frequency of hospitalizations as well as the mortality rate.

From a technology perspective, implementation of genomic and personalized medicine in Greece could facilitate the adopting of novel genotyping approaches in a clinical setting. Such technologies include next-generation sequencing and companion technologies such as array-on-demand diagnostics approaches for genetic diseases and non-invasive prenatal diagnosis. The interviewed stakeholders also believed that implementation of genomic and personalized medicine would lead to rapid development of the genomics and biotechnology industry in Greece. This would create opportunities for start-ups to be established that provide a more multidisciplinary product or service (e.g., combining innovative technology with analysis software).

Finally, stakeholders noted that implementing genomic and personalized medicine in Greece could strengthen national and international collaborations between corporate and academic key players in the genomic medicine field.

In terms of obstacles and threats that could hold the field back, stakeholders highlighted the low genetic awareness of the general public combined with the poor and/or incomplete genomics knowledge of healthcare professionals. If combined with a limited evidence base and a lack of tools to assist physicians with genomics-informed therapeutic decisions, this obstacle could slow implementation and deprive the public of the benefits of genome-guided treatment modalities.

Legislative challenges were also identified. First, Greek law does not currently guarantee personal data protection. To this end, stakeholders believe that insurance companies could deny insurance to patients based on their genetic profile, which leads to patient stigmatization. Healthcare reforms to rectify this deficiency are likely to elicit opposition from well-organized interest groups [19]. Legislation to control the provision of genetic services via direct-to-consumer or over-the-counter approaches is also limited, especially for tests that lack scientific evidence, such as nutrigenomic tests and genetic tests for multifactorial genetic diseases with strong environmental components. This presents a particular challenge when combined with the lack of genetic awareness from the public, as patients may not be able to distinguish between scientifically sound genetic tests and tests that lack a solid evidence base [14,17]. If the lack of a stable healthcare environment and of a consistent national strategy for genetics and genomics is added to the equation, then a serious threat arises that could hold back the field and weaken the contribution of genomic medicine interventions in the eyes of the main beneficiaries, which are the patients and the treating physicians.

Last, stakeholders considered the lack of funding for capacity building and limited reimbursement of genetic tests from the public and private insurance companies to be an obstacle of equal importance for the implementation of genomic medicine (see also Chapter 10).

9.6 Concluding remarks

In this chapter, we have outlined how stakeholder analysis can be undertaken to describe a policy environment and identify the role, views, interests, and positions of key stakeholders towards a given policy. We also summarized the findings from our previous work implementing stakeholder analysis to assess the views of key stakeholders related to genomic and personalized medicine in Greece.

This work revealed that most of the key stakeholders view the implementation of genomic and personalized medicine in Greece favorably, with just a few key stakeholders, such as the Ministry of Health and the public health insurance funds, currently opposing this new trend in medical practice. We anticipate that when robust evidence emerges on the effectiveness and cost-effectiveness of implementing personalized medicine interventions, the views of these key stakeholders will become more favorable. Also, these data underline the fact that most of the stakeholders seem to place their financial rather than self-interest at a higher priority. Most of the professional key players also express their scientific and professional interest, while the consumers highly prioritize their self-interest to access high-quality and affordable health services.

The stakeholder analysis that we have presented is currently being replicated in Greece to acquire further insights into how the views of stakeholders develop over time, within the scope of the Genome of Greece initiative (www.gogreece.org.gr). This analysis could also be replicated in other countries to understand how views and attitudes differ by country [20], which will allow for the development of harmonized national policies to implement genomic and personalized medicine in clinical practice. This would not only expedite implementation but also catalyze reimbursement of genomic and personalized medicine interventions (see Chapter 10).

Acknowledgments

This work was partly funded by the University of Patras research budget and a Golden Helix Foundation research grant to GPP. This study was encouraged by the Genomic Medicine Alliance Public Health Genomics Working Group. GPP is Full Member and National Representative of the European Medicines Agency, Committee for Human Medicinal Products (CHMP)—Pharmacogenomics Working Party, Amsterdam, the Netherlands.

References

[1] Squassina A, Artac M, Manolopoulos VG, Karkabouna S, Lappa-Manakou C, Mitropoulos K, Manchia M, del Zompo M, Patrinos GP. Translation of genetic knowledge into clinical practice: the expectations and realities of pharmacogenomics and personalized medicine. Pharmacogenomics 2010;11(8):1149–67.

[2] Cooper DN, Chen JM, Ball EV, et al. Genes, mutations, and human inherited disease at the dawn of the age of personalized genomics. Hum Mutat 2010;31(6):631–55.

[3] Reydon TA, Kampourakis K, Patrinos GP. Genetics, genomics and society: the responsibilities of scientists for science communication and education. Perinat Med 2012;9(6):633–43.

[4] Hietala M, Hakonen A, Aro AR, Niemelä P, Peltonen L, Aula P. Attitudes toward genetic testing among the general population and relatives of patients with a severe genetic disease: a survey from Finland. Am J Hum Genet 1995;56(6):1493–500.

[5] Mai Y, Koromila K, Sagia A, et al. A critical view of the general public's awareness and physicians' opinion of the trends and potential pitfalls of genetic testing in Greece. Perinat Med 2011;8(5):551–61.

[6] Mai Y, Mitropoulou C, Papadopoulou XE, Vozikis A, Cooper DN, van Schaik RH, Patrinos GP. Critical appraisal of the views of healthcare professionals with respect to pharmacogenomics and personalized medicine in Greece. Perinat Med 2014;11(1):15–26.

[7] Makeeva OA, Markova VV, Roses AD, Puzyrev VP. An epidemiologic-based survey of public attitudes towards predictive genetic testing in Russia. Perinat Med 2010;7(3):291–300.

[8] Pavlidis C, Karamitri A, Barakou E, Cooper DN, Poulas K, Topouzis S, Patrinos GP. Analysis and critical assessment of the views of the general public and healthcare professionals on nutrigenomics in Greece. Perinat Med 2012;9(2):201–10.

[9] Kampourakis K, Vayena E, Mitropoulou C, van Schaik RH, Cooper DN, Borg J, Patrinos GP. Key challenges for next-generation pharmacogenomics. EMBO Rep 2014;15(5):472–6.

[10] Reich RM. Applied political analysis for health policy reform. Curr Issues Public Health 1996;2:186–91.

[11] Reich MR, Cooper DM. PolicyMaker: computer-assisted political analysis. Software and manual. Newton Centre, MA: PoliMap; 1996.

[12] Mintzberg H. The rise and fall of strategic planning. New York: Free Press; 1994.

[13] Sagia A, Cooper DN, Poulas K, Stathakopoulos V, Patrinos GP. A critical appraisal of the private genetic and pharmacogenomic testing environment in Greece. Perinat Med 2011;8(4):413–20.

[14] Patrinos GP, Baker DJ, Al-Mulla F, Vasiliou V, Cooper DN. Genetic tests obtainable through pharmacies: the good, the bad, and the ugly. Hum Genom 2013;2013(7):17.

[15] van der Wouden CH, Cambon-Thomsen A, Cecchin E, Cheung KC, Dávila-Fajardo CL, Deneer VH, Dolžan V, Ingelman-Sundberg M, Jönsson S, Karlsson MO, Kriek M, Mitropoulou C, Patrinos GP, Pirmohamed M, Samwald M, Schaeffeler E, Schwab M, Steinberger D, Stingl J, Sunder-Plassmann G, Toffoli G, Turner RM, van Rhenen MH, Swen JJ, Guchelaar HJ, Ubiquitous Pharmacogenomics Consortium. Implementing pharmacogenomics in Europe: design and implementation strategy of the Ubiquitous Pharmacogenomics Consortium. Clin Pharmacol Ther 2017;101(3):341–58.
[16] European Commission (EC). The second economic adjustment programme for Greece: third review. In: European Economy, occasional papers 159, European Commission, Directorate-General for Economic and Financial Affairs, Brussels; 2013. Publications.
[17] Kechagia S, Yuan M, Vidalis T, Patrinos GP, Vayena E. Personal genomics in Greece: an overview of available direct-to-consumer genomic services and the relevant legal framework. Public Health Genom 2014;17 (5-6):299–305.
[18] Patrinos GP, Mitropoulou C. Measuring the value of pharmacogenomics evidence. Clin Pharmacol Ther 2017;102(5):739–41.
[19] Mitropoulou C, Mai Y, van Schaik RH, Vozikis A, Patrinos GP. Documentation and analysis of the policy environment and key stakeholders in pharmacogenomics and genomic medicine in Greece. Public Health Genom 2014;17(5-6):280–6.
[20] Mitropoulos K, Cooper DN, Mitropoulou C, Agathos S, Reichardt JKV, Al-Maskari F, et al. Genomic medicine without borders: which strategies should developing countries employ to invest in precision medicine? A new "fast-second winner strategy. OMICS 2017;21(11):647–57.

CHAPTER

10 Feasibility for pricing, budget allocation, and reimbursement of personalized medicine interventions

Christina Mitropoulou[a,b], Margarita-Ioanna Koufaki[a], Athanassios Vozikis[c], and George P. Patrinos[b,d,e]

[a]The Golden Helix Foundation, London, United Kingdom, [b]Department of Genetics and Genomics, College of Medicine and Health Sciences, United Arab Emirates University, Al-Ain, United Arab Emirates, [c]Economics Department, University of Piraeus, Piraeus, Greece, [d]Department of Pharmacy, School of Health Sciences, University of Patras, Patras, Greece, [e]Zayed Center of Health Sciences, United Arab Emirates University, Al-Ain, United Arab Emirates

10.1 Introduction

Personalized medicine could facilitate individualized drug treatment and earlier detection of disease onset, either pre-symptomatically or through the determination of individual risk, which could prevent disease progression. At present, a personalized medicine approach is possible in a small number of monogenic disorders (e.g., hemoglobinopathies, cystic fibrosis), hereditary cancers, and more than 200 of the commonest drug treatment modalities [1]. Personalized medicine has the potential to shift the emphasis in medicine from clinical and/or therapeutic intervention to prevention, inform the selection of optimal treatment modalities and reduce trial-and-error prescribing based on an individual's genome, help avoid adverse drug reactions, and improve quality of life. Personalized medicine can also reveal alternative or additional uses of existing medications (drug repositioning) and direct the selection and design of novel therapeutics. Overall, personalized medicine interventions and approaches could aid in containing the cost of health care in the medium-to-long term [2].

However, economic and policymaking issues present a serious hurdle to realizing the benefits of personalized medicine [3]. As described in Chapter 9, multiple stakeholders in society play a role in the implementation of personalized medicine, including research organizations and academic centers, corporate entities, funders and payers, patients and the general public, innovators, regulators, policymakers, and legislators. Previous work suggests that many of these stakeholders are generally favorable to the implementation of personalized medicine [4–6]. As such, if

Economic Evaluation in Genomic and Precision Medicine. https://doi.org/10.1016/B978-0-12-813382-8.00010-0

policies and measures are introduced to foster the overall positive attitude of most stakeholders, ensuring that personalized medicine interventions are implemented in an efficient and effective manner, the remaining stakeholders (who hold neutral-to-negative opinions) may also be supportive [4].

The most important factors required for the successful implementation of personalized medicine interventions into the public healthcare system include:

(a) political engagement and willingness to change existing healthcare systems and processes,
(b) high-quality scientific evidence demonstrating clinical efficacy and safety,
(c) evidence for acceptability, fairness, solidarity, and appropriateness,
(d) appropriate policy measures and legislation,
(e) evidence on affordability and cost-effectiveness, and
(f) appropriate knowledge and education for clinicians and patients [7–9].

If one or more of the preceding factors is not met, smooth implementation of personalized medicine may be hampered; as illustrated in Fig. 10.1, there is a missing piece in the jigsaw puzzle. For example, if payers and insurance companies are unwilling to reimburse the costs of personalized medicine interventions, progress towards incorporation and implementation will be stalled. Reimbursement decisions in genomic testing are complicated and although genomic testing has been performed for more than 15 years, the respective decision-making process is still evolving, as discussed in Sections 10.2 and 10.3 [10,11].

In the following paragraphs, we present approaches and policies that are pursued for pricing and reimbursement in different countries and adopted by the respective regulatory bodies and health technology agencies (HTAs), as well as define the prerequisites that the national pricing and reimbursement strategies should fulfill for personalized medicine interventions.

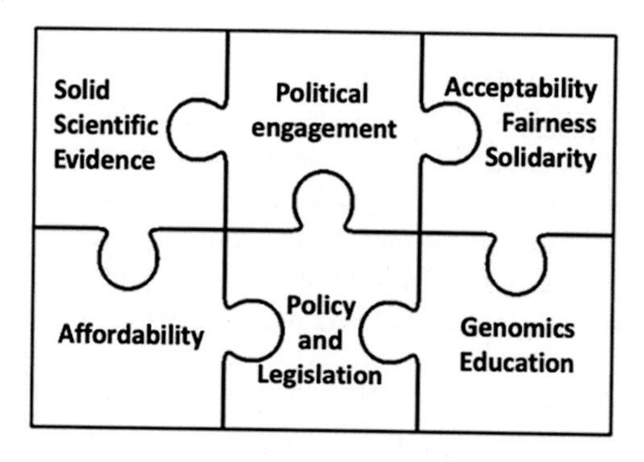

FIG. 10.1

Conceptual depiction of the interdependencies between different political, societal, and economic factors affecting pricing and reimbursement of personalized medicine interventions as the interconnected pieces of a jigsaw puzzle. If one of the pieces of the puzzle is missing, most importantly cost-effectiveness and affordability, then progress towards incorporation and implementation of personalized medicine interventions will be stalled (see text for details).

10.2 Institutions involved in pricing and reimbursement

Personalized medicine interventions are classified as a sub-category of medical devices in regulatory agencies. Regulation of medicinal products along with the medical device environment varies considerably between countries across the globe. In the European Union, all member states follow the EU directive on medical device regulations. Within this directive, a medical device must have the Conformité Européenne (CE) mark to be sold within the respective member countries [12]. EU member states have also implemented complex reimbursement policies, requiring medical devices to be included on approved reimbursement lists. Aside from the common EU directives, the regulatory structure for the reimbursement of medical devices, including personalized medicine interventions, differs across the members, and each member adopts a different pricing and reimbursement approach. Indeed, there are different pricing and reimbursement institutions in each country, which constitutes a particular form of autonomous governmental bodies with regulatory or advisory jurisdiction who collaborate closely with HTAs [13].

For instance, reimbursement is organized via Der Gemeinsame Bundesausschuss (G-BA) in Germany, La Haute Autorité de Santé (HAS) in France, the National Health Service (NHS) in the United Kingdom, Il Servizio Sanitario Nazionale (SSN) in Italy, and El Instituto Nacional de la Salud (INS) in Spain. Even if the available data source (clinical evidence from clinical trials, epidemiological data, statistics) provided by pharmaceutical companies are common, each agency considers a different set of parameters and evaluates them in a different way. Consequently, reimbursement for specific personalized medicine interventions demonstrates great heterogeneity across the European Union due to differences in national priorities, healthcare budget constraints, healthcare policies, and so on [13].

A typical example of this is the case of Belgium in which there are dual track. More precisely, the reimbursement of diagnostic-related devices such as genomic testing is claimed by the producer (pharmaceutical company) who follows a specific process regulated by the National Institute for Health and Disability Insurance, while the national HTA (KCE) is possible to select some of the candidates to undergo a full evaluation by HTA.

In the United States, the Centers for Medicare and Medicaid Services (CMS) manage the Clinical Laboratory Improvement Amendments (CLIA) program, which regulates clinical laboratories, including those performing genomic testing. Several government agencies in the United States are currently working towards the development of regulatory standards and accreditations for genomic testing laboratories to establish a comprehensive integration of genomic testing into clinical practice [14]. Furthermore, the Centers for Disease Control and Prevention (CDC) has a genomic testing policy group specifically focused on the CLIA regulations, as well as projects studying the validation of genomic tests and their integration into clinical practice. In addition, the Secretary's Advisory Committee on Genetics, Health, and Society (SACGHS) issued a comprehensive report summarizing the issues surrounding the reimbursement of genomic tests [9] (see next paragraph).

The Center for Medical Technology Policy (CMTP) and the center of Evaluation of Genomic Applications in Practice and Prevention (EGAPP) have launched several programs with the objective to resolve all issues related to reimbursement of genomic testing; they methodically work on setting objective reimbursement criteria. These efforts aim to make pricing decisions more transparent and to facilitate decision-making within the pharmaceutical industry in terms of developing innovative products [15]. Finally, the US National Human Genome Research Institute (NHGRI) advances personalized medicine by assisting payers to evaluate emerging genomic tests for reimbursement and by promoting research into demonstrating health benefits and cost-effectiveness of personalized medicine interventions.

Another interesting case is the 3-I framework, which stands for Ideas, Interests, and Institutions, which are the three main pillars of policymaking. It is a technique commonly applied by many policymakers and politicians worldwide mainly in the social security and welfare domain. After 1996, the 3-I framework was successfully applied to the healthcare sector for the treatment of AIDS, but the most recent example is its use in Canada. Canada has one of the most prominent and successful genome sequencing programs, known as Genome Canada. The Canadian government along with Canadian HTA CADTH, Health Canada, is working to establish a centralized pricing approach for personalized medicine technologies and interventions, a fact that has been emphasized several times by national laboratories and scientists [16]. Nowadays, the Canadian government has applied the 3-I framework to deal with the existing challenges in pricing and reimbursement of personalized medicine applications.

Finally, there are great examples of institutions involved in pricing from Latin American countries where there are several ongoing genome sequencing programs and collaborations across countries. More precisely, RedETSA is an HTA network of 17 countries represented by 34 different national institutions such as research institutions, ministries of health, regulatory authorities, HTAs, collaborating centers, and non-profit organizations of the Pan American Health Organization/World Health Organization (PAHO/WHO). This initiative was funded in 2011 and supported by WHO with the aim to improve access to HTA information and practices, harmonize the approval and reimbursement guidelines of health technologies, and strengthen the sustainability of health systems in low-income countries in the Americas. Red de Evaluación de Tecnologias en Salud de las Américas (RedETSA) along with the Red Iberoamericana Ministerial de Aprendizaje e Investigación en Salud network (RIMAIS), ISPOR Latin America, and national HTAs such as CENETEC in Mexico, CONITEC in Brazil, ITES in Colombia, IECS in Argentina, and HAD in Uruguay work on participating in personalized medicine-related programs, familiarizing the public, governors, healthcare professionals, and policymakers with personalized medicine technologies, and reimbursing personalized medicine interventions. These organizations have already established one of the biggest collaborations in the healthcare sector, and they constitute an excellent example of how international partnerships can reinforce adoption of personalized medicine.

To conclude, it is evident that approval of a personalized medicine intervention from regulatory bodies as well as its pricing and reimbursement is quite complicated because the regulatory and pricing landscape varies among countries. These observed differences may provoke potential barriers in personalized medicine adoption and reimbursement by decreasing their rate of implementation.

10.3 Coverage, pricing, and reimbursement strategies for genomic testing services

National healthcare systems in Europe, Asia, the United States, and other countries worldwide are currently wrestling with various issues such as increasing healthcare costs, long patient waiting times, and more. The situation is getting more complicated with the introduction of new pioneering healthcare technologies and interventions such as personalized medicine technologies [15]. Since the appearance and expansion of the personalized medicine approach, there have been different and convergent opinions and perceptions of such an innovation. Apart from the evident healthcare benefits in drug and disease management, personalized medicine is a breakthrough that can potentially revolutionize patient management overall.

This aspect of personalized medicine has sparked a controversy related to potential harms and risks related to bioethics, data protection and privacy, and of course costs [14] The inexistence of a common regulation, the ambiguity in clinical implementation, and the serious lack of specific evidence-based criteria to price and evaluate such technologies lead to delays. This situation has posed a serious challenge in the reimbursement of innovative and often expensive molecular diagnostic tests and genomic testing services since they have a considerably high impact on healthcare budget.

For this reason, different approaches are adopted by regulatory agencies and HTAs. The most common approach is the implementation of economic evaluation theories and tools such as cost-effectiveness and cost-utility analysis to provide stakeholders with official economic data. While evidence on cost-effectiveness gives valuable information, from the payers' perspective, this information may or may not be considered in their decision-making process (it depends on the approach used; see Section 10.2). Achieving direct cost savings is more crucial for payers, particularly in the case of personalized medicine interventions, owing to the fact that it is possible to avoid adverse drug reactions, delay the onset of disease-related complications, and reduce the time required for a patient to respond to a therapeutic scheme [12,13].

Indeed, the European reimbursement environment is not unified, as each member state has its own policies and reimbursement is approved by either private insurance companies or public payer(s) or, in some cases, a combination of the two. Approval for reimbursement from public health providers often requires lengthy negotiations. Most European countries review and authorize tests at a local level, not a national one, which can create a significant hurdle to consistent market access for genomic tests and personalized medicine services [12].

In the United States, there is another pricing approach implemented. More precisely, the US Food and Drug Administration (FDA) has implemented the Current Procedural Terminology (CPT) code since 2005 for the reimbursement of personalized medicine interventions and genomic testing (such as PGx), which have already been launched into the market [17]. This coding system is updated annually and enables healthcare practitioners to accurately document any medical, surgical, and diagnostic procedures or services offered and report them to either governmental, public, or private payers depending on the state's requirements. The CPT code initiative is a great pricing model for medical devices, but it can be quite restricting for personalized medicine interventions due to their uniqueness and peculiarities.

In Latin American countries, the EULAC-PerMed program intends to familiarize local HTAs and policymakers with personalized medicine interventions to encourage their adoption by securing international collaborations and partnerships and by setting specific directives on how to price and reimburse such technologies (www.eulac-permed.eu/). The program was an EU initiative started in 2019 and it includes six Latin American countries and their HTAs. Although EULAC-PerMed is using the economic methods followed by EU members, it has gained acceptance from local governments because it respects cultural differences and it incorporates the social norms, economic status, and political environment of each country.

In conclusion, it is considered that each country will adopt different approaches for pricing and reimbursing innovative technologies such as personalized medicine. This variance complicates the adoption of these approaches and it is still unclear which is the best practice to apply in terms of reimbursement for personalized medicine.

10.4 A proposed strategy for pricing and reimbursement in personalized medicine

In the previous sections, we described national approaches applied in the reimbursement of personalized medicine interventions, including genetic testing services for the detection of early disease onset and/or disease progression and individualized drug treatment modalities. In this section, we define the basic characteristics of a successful national strategy for pricing, budget allocation, and reimbursement for personalized medicine interventions.

There are four main prerequisites that a national strategy should satisfy:

(1) Ensure access to essential genomic testing services for all and at acceptable and affordable prices for the healthcare system.

(2) Establish a common and centralized regulation at a national level to ensure safety, efficacy, quality, fairness, and solidarity, while allowing space for innovation necessary to move the field forward.

(3) Ensure the proper implementation of personalized medicine interventions and evaluate information by physicians according to patient needs and clinical utility/actionability of testing outcomes.

(4) Invest in the research of personalized medicine, evaluate novel and existing diagnostic procedures, and monitor patient safety.

10.4.1 Ensure access to essential genomic testing services for all at acceptable and affordable prices for the healthcare system

Essential genetic tests are those tests that satisfy the priority healthcare needs of a given population. Like essential medicines, these tests are selected with regard to disease prevalence, public health relevance, evidence of clinical efficacy and safety, and comparative costs and cost-effectiveness. Examples of essential genetic tests include pharmacogenomics (PGx) tests, and genetic tests for monogenic diseases (e.g., β-thalassemia, cystic fibrosis) and hereditary cancers (e.g., BRCA1/2-associated breast cancer).

This criterion derives from the theory of the "iron triangle" in the healthcare sector and the need for everyone's accessibility to health facilities, health technologies, medications, and general healthcare treatment. This setting isn't always feasible because it is necessary to meet the aspect of cost. Personalized medicine interventions such as genomic testing are still rather expensive for a healthcare system to cover, but this cost dramatically decreases as the rate of personalized medicine adoption increases.

Genomic testing should be available for all citizens in a timely manner and be distributed according to each person's therapeutic needs, no matter the person's locality, age, gender, or income. The adoption of this approach is believed to improve health outcomes and increase the implementation of personalized medicine in clinical practice. Currently, genetic testing is becoming more and more popular among clinicians who prescribe a wide range of testing. Many studies have described a conflicting and perplexing reality regarding coverage of clinically relevant genetic testing. It seems that most of those tests aren't covered or reimbursed by public and/or private insurance schemes in many countries. The main reason is the reference indication and the clinical utility of the test for each subject. In other words, the percentage of coverage of each test depends on whether this genetic testing is preemptive or reactive [18]. Indeed, in many countries, including the United States, the percentage of testing coverage and its reimbursement is mainly determined by the medical indication and prescription reason [19]. If the subject is a patient with a diagnosed disease, the genetic intervention will be fully covered, whereas if the testing aim is investigational, the test is not fully covered by the subject's insurance scheme [19].

However, there are several free genetic screening programs available in many countries, including Australia, Qatar, the Netherlands, France, and others. For instance, in the Netherlands, citizens are invited to undergo at least one genetic test for free during their lives for both preemptive and reactive reasons [20]. In Australia, certain regions have implemented a listing of genetic testing for a series of childhood syndromes and intellectual disability that is subsidized by the federal government and available to groups of people with an increasing genetic disposition to develop such disorders [21].

Therefore, undergoing genetic testing could depend on the patient's ability to pay rather than the clinical utility, need, therapeutic rationale, or cost-effectiveness of the testing [22]. In an emerging third-party insurance reimbursement landscape, patient self-pay is an option, but only for patients with discretionary funds, which is a serious barrier for the widespread adoption of genomic testing in clinical practice.

To deal with patient discrimination, it would be beneficial to define and select the most often used essential genomic tests. This process would contribute to more rational prescribing, less confusion, and greater familiarity with genomic testing among diagnostic laboratories and healthcare providers, provided of course that this process is regularly updated by clinical guidelines [23]. Classification of every genomic test should be supported by cost-effectiveness analyses. The key stakeholders tasked with ensuring the availability and affordability of genomic tests are ministries of health, regulatory bodies such as the national medicines authorities, payer organizations, medical device manufacturers, academic and other research institutes, pharmaceutical companies, pharmacies, hospitals and clinics, public and private diagnostic laboratories, and physicians [4,23].

At the same time, making personalized medicine interventions affordable is radical to ensure population access to essential genomic tests that are supported by high-quality scientific evidence (see also Fig. 10.1). Increasing affordability should be a key commitment of governments to guarantee that pecuniary issues do not create barriers to patients' access to genomic testing. Also, an effective, transparent, sustainable, and robust pricing system should be in place where different countries negotiate prices to their own healthcare setting. Under such an arrangement, pricing decisions would consider the comparative effectiveness of new genomic tests and their incremental effects relative to their incremental costs as components of a cost-effectiveness analysis for the new personalized medicine interventions only when relevant data on health outcomes are available.

A first step would be the formation of a pricing committee within a country's HTA or regulatory authority that would oversee the pricing of all genomic tests. Members of the pricing committee should conduct regular, timely, and robust background checks and maintain databases in order to advise appropriate prices for all new (or existing) genomic tests entering the market, aiming to allow reimbursement by public insurance funds. For certain genomic tests, a tendering system might be developed to promote price competition between manufacturers or, most importantly, distributors. However, such a system should be well regulated and ultimately transparent.

The criteria of genomic testing classification should also be carefully controlled and regularly reviewed, especially in relation to the emergence of clinical, economic, fiscal, or other criteria, and how these different criteria should be weighed against each other in the decision-making process.

In addition, a separate list of genomic tests that fail to meet certain scientific standards and/or lack clinical utility, such as genetic tests for athletic performance, nutrigenomic tests, intelligence criminality, personality, genomic identity testing, and others, should also be established [24,25] for public awareness and information.

This proposal seems to be difficult to implement since the available clinical outcomes and cost-effectiveness data are still limited, but it is possible to work in the future. Nevertheless, it is highly important that each country works towards this orientation to achieve its goals.

10.4.2 Establish a common and centralized regulation to ensure safety, efficacy, quality, and fairness, while allowing space for innovation necessary to move the field forward

Robust regulation is a key to ensure the safety, efficacy, and quality of genomic tests and should be a result of strong cooperation among all relevant stakeholders. According to Section 10.1, existing regulations and legislations related to medical device reimbursement vary across countries at local, subregional, and regional levels, and they are sometimes lengthy and not so flexible to innovation. This non-harmonized setting causes a series of problems for policymakers and healthcare stakeholders because it raises many potential weaknesses and threats to be addressed.

The current situation could be ameliorated thanks to the development of common evaluation criteria, both clinical and economic, to ensure the quality and transparency of procedures and to preserve the opportunity for innovation [26]. Safety assurance is extremely important for regulatory bodies' decision-making and thus a safety process must be enacted for personalized medicine interventions to address issues pertaining to credibility, accuracy, and/or testing reliability.

10.4.3 Implementation of genomic tests and information by physicians, according to patient needs and clinical utility/actionability of testing outcomes

According to the literature, one of the main barriers that impedes the adoption of innovative healthcare technologies in the clinical setting is the low level of awareness and expertise of healthcare professionals [14,26]. Healthcare professionals play a dominant role in the implementation of new technologies in health, as they are the end users of the applications and any false or inaccurate application of such interventions may negatively impact on the patient and their family members. Healthcare professionals include clinicians and/or pharmacists, depending on the country and national regulations (e.g., pharmacists in the Netherlands can prescribe drugs).

Indeed, physicians are responsible for choosing the appropriate personalized medicine intervention for their patients, prescribing it, and monitoring their patients' progress. Having well-informed and specialized personnel who acknowledge the clinical significance of the technology who can apply it effectively is an essential step towards securing its appropriate use in clinical routine. In personalized medicine, physicians and patients act in concert to achieve the optimum outcome. As such, patients should be aware of the benefits and risks of such technologies and understand their importance for early disease diagnosis and treatment.

Furthermore, physicians' opinions should always be taken into consideration, as they constitute a valuable source of real-world data necessary for decision-making. It would be desirable to conduct surveys among clinicians to establish if there are any concerns about companion diagnostic products or monitoring tools. As such, development of an electronic prescribing system could be essential for more effective and appropriate prescribing of genomic tests. Such a system would be designed to monitor physicians' prescribing, providers' implementation, and pharmacists' distribution records to rationalize the provision of genomic tests. A good example is the use of disease-specific "Clinical Utility Gene Cards," which establish a priori peer-reviewed criteria for indication of genomic testing [27]. Electronic prescribing systems are gradually being adopted by several healthcare systems to reduce healthcare costs and maintain the quality of health care provided to patients, especially when coupled to electronic health records [28].

Lastly, an important parameter in the prescription of genomic tests to patients and the public in general is their provision according to international classification and existing guidelines, especially in the case of PGx testing, which depends upon standardized prescribing protocols across therapeutic categories. This underlines the need to provide personalized medicine interventions to patients at the level of healthcare professionals, namely, the physician and the pharmacist, and not using the direct-to-consumer model, in which case the healthcare professional part is missing from the equation and negatively impacts on the proper provision of healthcare services to the patient.

10.4.4 Invest in the research of personalized medicine, evaluate novel and existing diagnostic procedures, and monitor patient safety

In the era of technological advancement, investing in pioneering interventions seems to be the only way to progress. Personalized medicine interventions are a promising technology that should be invested in. Research is a critical component for the personalized medicine sector and is required for newer, innovative approaches and products. As previously mentioned, there is an urgent need for clinical evidence to support decision-making processes and improve pricing evaluation. Due to the pioneering nature of personalized medicine interventions, policymakers and regulatory agencies seek evidence of clinical efficacy and effectiveness to overcome existing drawbacks and economic burdens. Thus, research not only has the potential to save money and reduce healthcare expenditures but also to contribute to the economy as a whole through the development of innovative, patentable, and profitable technologies, tools, and services.

Investment in research and development is partly dependent on the healthcare system, the national priorities for healthcare research, and on a variety of factors both financial and non-financial. National policy, for example, is a key factor as it may promote a more flexible and responsive regulatory framework to attract clinical trials and might encourage investment in research and development in personalized medicine. Also, HTA activities can contribute significantly to this direction, for example, by contributing to the creating and updating of genomic test lists by payers, specifying clinical and prescribing protocols based on clinical cost-effectiveness criteria, and advancing e-health systems.

Apart from the investment in research and development, it is important to give incentives that have a major impact on expediting integration of personalized medicine services clinically, with would reciprocally direct impact on public health. Investing in human capital could be a successful strategy for the wider implementation of personalized medicine interventions. For instance, funding continuous education, such as genomics education, of healthcare professionals and providers would increase genomic literacy and ensure the proper evaluation of novel and existing diagnostic procedures [29,30]. Furthermore, it would be important to hire allied professionals such as genetic counselors who are well-trained specialists in cutting-edge technologies because they would speed up the reimbursement process of such technologies while increasing their objectivity.

Finally, money needs to be invested to develop new tools such as translational tools that could revolutionize some therapeutic areas or next-generation sequencing-based panels and approaches. New methods and practices would increase the available scientific data, and they can be used for monitoring and safety assessment purposes since patient wellbeing, safety, and health are the main priorities of national policies and regulatory agencies.

10.5 Restrictions and concerns

Although lower prices may emerge from the development of increasingly efficient and widespread technologies and interventions in personalized medicine, it is crucial that reimbursement strategies not only ensure broad access to high-quality interventions but also continue to encourage the development of a pipeline of more innovative interventions requiring substantial risk-based research. There are many barriers and concerns that affect and subsequently restrict the reimbursement rate of testing.

At first, many genomic tests, particularly those for predicting complex diseases and phenotypes, require more data derived from randomized controlled clinical trials as well as prospective clinical trials to enhance clinical utility and efficacy. As such, these tests are currently not subsidized or reimbursed [14,31]. This fact is aligned with the main concerns of clinicians about the implementation of personalized medicine interventions in clinical practice. In many studies, it is highlighted that healthcare professionals are not confident with their level of knowledge and training in genomics and other personalized medicine applications, while they pinpoint the inexistence of unified clinical guidelines [32,33]. Moreover, another common issue is the lack of interpretation or "translation" tool that would assist and support physicians' decision-making [34].

Finally, there are also some societal aspects that need to be considered. Cultural and religious norms may affect the adoption of innovative products and services and pose a significant burden for their reimbursement. The public's concerns about data privacy and confidentiality may also adversely affect the adoption of personalized medicine. Admittedly, the existing legislation about the use and storage of genomic or other clinical data is quite vague and in some cases absent, whereas there are no common directives and laws enacted by international institutions except for the example of the European Union's General Data Protection Regulation (GDPR) guidelines. This legal gap could pose a serious threat for the progress of personalized medicine coverage [26].

Given that molecular diagnostics have no direct health improvement effect but potentially impact on generating downstream health effects, all the aforementioned concerns have a great impact on the perceptions/attitudes of healthcare stakeholders and policymakers as well. Being reluctant about the usefulness of personalized medicine can affect the opinion of a stakeholder who is expected to make the right policy decisions in this area. This key decision affects the coverage and reimbursement of personalized medicine interventions and constitutes the biggest challenge for all stakeholders.

Restricting a patient's access to testing due to restricted coverage, since laboratories tend to offer only those tests that are mostly reimbursed contrary to those that are paid by the patients themselves, is an important disadvantage. This requires better regulatory measures to minimize the impact of the "free-market" shortcomings in the healthcare sector while incentivizing investments for research and development for genomic tests and companion diagnostic solutions [23]. This would entail better coordination of the various HTA processes within a country specifically aimed at companion diagnostic solutions.

As mentioned, HTAs represent a rapidly growing area of government policy, especially over the last decade that governments seek to contain costs. However, most of the paradigms for healthcare appraisal and technology assessment have been developed for the purpose of comparing procedures or drug therapies. Although most stakeholders would agree that (pharmaco)genomic tests are beneficial by having an impact both on patient management and the more accurate delivery of drug treatment modalities, many are still concerned that processes underlying "clinical utility" assessments are often not clear [6,35].

To this end, more objective and reliable standards for these HTA and economic evaluation processes need to become broadly accepted. Since many argue that reimbursement should be tied to value, that is, the actual «*value*» that includes not only the economies for the payer but, most importantly, the utility for the patient, needs to be acceptably defined and prioritized, taking into consideration how difficult it has been to consider all factors involved in «*value*» and to implement value-based pricing [2].

As such, improving genomic testing quality reduces the rate of incorrect results (i.e., false positive or false negative). Test quality can also be ensured by laboratory accreditation, which constitutes the most extensive quality endorsement, covering genomic test quality inside the laboratory and at the interface with clinicians. To this end, all laboratories performing molecular diagnostic tests should be accredited and existing external quality assessment schemes should be extended both at national and international levels. In fact, external quality assessment schemes have been shown to increase the accuracy of genomic diagnostic testing [36,37].

10.6 Conclusions

Personalized medicine presents special challenges to pricing and reimbursement because of the rapid evolution of new technologies and the lack of a clear understanding by all stakeholders of the requirements for inclusion in coverage policies. Coverage and reimbursement of genomic tests is complicated because no consensus yet exists as to what should be covered. Payers must therefore consider options such as bundled payments or risk-sharing agreements so that genomic tests are connected to specific drug treatment modalities and/or an entire continuum of health care. This could positively affect implementation, which would subsequently help to advance the field of genomic testing.

A well-defined pricing and reimbursement policy should play a central role in the overall pricing and reimbursement process, intersecting the delivery of an innovative personalized medicine intervention from conception to the end user (the patient) through well-educated and thoroughly informed clinicians and healthcare professionals (Fig. 10.2). HTAs are therefore of paramount importance, and mostly depends on economic evaluation of personalized medicine interventions [38].

To conclude, genome-informed targeted therapies are decisive in improving patient care and outcomes and in several places seem to be superior to the currently practiced "one-size-fits-all" approaches. These, including the already approved personalized medicine approaches, tailored to the patient's genotypic profile, are expected to be the most cost-effective approach to clinical practice, coupled with well-educated clinicians and healthcare professionals as well as "educated"

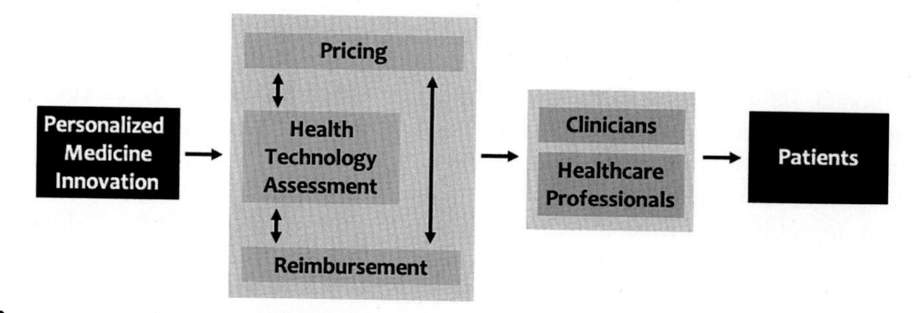

FIG. 10.2

Overview of the stages an innovative personalized medicine intervention (e.g., a molecular diagnostics test or companion diagnostic assay) needs to go through before it becomes accessible to patients and society. In this process, health technology assessment plays a pivotal role.

and well-aware patients. Personalized medicine will eventually become the standard of care and could in turn help to reduce healthcare expenditure [38].

In addition, model-based cost-effectiveness analysis (see Chapter 3) is a useful tool that, as in the United States and most countries in Europe, must be embedded into the mainstream HTA processes to provide information on the incremental costs of new interventions such as next-generation sequencing technologies [39]. Robust studies that span across all core components of genomic testing services (i.e., technology, diagnostic components, and service delivery) are needed to complement cost-effectiveness analyses to inform decisions for reimbursement of personalized medicine interventions.

Lastly, adoption of an appropriate legal framework is necessary to determine the appropriate conditions for reimbursement of clinically valid tests and to set the foundations of a stable, effective, and transparent pricing system.

References

[1] Manolio TA, Abramowicz M, Al-Mulla F, Anderson W, Balling R, Berger AC, Bleyl S, Chakravarti A, Chantratita W, Chisholm RL, Dissanayake VH, Dunn M, Dzau VJ, Han BG, Hubbard T, Kolbe A, Korf B, Kubo M, Lasko P, Leego E, Mahasirimongkol S, Majumdar PP, Matthijs G, McLeod HL, Metspalu A, Meulien P, Miyano S, Naparstek Y, O'Rourke PP, Patrinos GP, Rehm HL, Relling MV, Rennert G, Rodriguez LL, Roden DM, Shuldiner AR, Sinha S, Tan P, Ulfendahl M, Ward R, Williams MS, Wong JE, Green ED, Ginsburg GS. Global implementation of genomic medicine: we are not alone. Sci Transl Med 2015;7(290). 290ps13.

[2] Patrinos GP, Mitropoulou C. Measuring the value of pharmacogenomics evidence. Clin Pharmacol Ther 2017;102(5):739–41.

[3] Snyder SR, Mitropoulou C, Patrinos GP, Williams MS. Economic evaluation of pharmacogenomics: a value-based approach to pragmatic decision making in the face of complexity. Public Health Genom 2014;17(5-6):256–64.

[4] Mitropoulou C, Mai Y, van Schaik RH, Vozikis A, Patrinos GP. Stakeholder analysis in pharmacogenomics and genomic medicine in Greece. Public Health Genom 2014;17(5-6):280–6.

[5] Mai Y, Koromila T, Sagia A, Cooper DN, Vlachopoulos G, Lagoumintzis G, Kollia P, Poulas K, Stathakopoulos V, Patrinos GP. A critical view of the general public's awareness and physicians' opinion of the trends and potential pitfalls of genomic testing in Greece. Perinat Med 2011;8(5):551–61.

[6] Mai Y, Mitropoulou C, Papadopoulou XE, Vozikis A, Cooper DN, van Schaik RH, Patrinos GP. Critical appraisal of the views of healthcare professionals with respect to pharmacogenomics and personalized medicine in Greece. Perinat Med 2014;11(1):15–26.
[7] Mette L, Mitropoulos K, Vozikis A, Patrinos GP. Pharmacogenomics and public health: implementing "populationalized" medicine. Pharmacogenomics 2012;13(7):803–10.
[8] Prainsack B. Personhood and solidarity: what kind of personalized medicine do we want? Perinat Med 2014;11(7):651–7.
[9] Williams MS. The public health genomics translation gap: what we don't have and why it matters. Public Health Genom 2012;15(3-4):132–8.
[10] Williams MS. The Genetic Future: Can genomics deliver on the promise of improved outcomes and reduced costs? Background and recommendations for health insurers. Disease Manag Health Outcomes 2003;11 (5):277–90.
[11] Williams MS. Insurance coverage for pharmacogenomic testing. Perinat Med 2007;4(4):479–87.
[12] Merchant M. Pricing and reimbursement strategies for diagnostics: Overcoming reimbursement issues and navigating the regulatory environment. Warwick, UK: Business Insights Ltd.; 2010.
[13] Angelis A, Lange A, Kanavos P. Using health technology assessment to assess the value of new medicines: results of a systematic review and expert consultation across eight European countries. Eur J Health Econ 2018;19(1):123–52. https://doi.org/10.1007/s10198-017-0871-0.
[14] Klein ME, Parvez MM, Shin JG. Clinical implementation of pharmacogenomics for personalized precision medicine: barriers and solutions. J Pharm Sci 2017;106(9):2368–79. https://doi.org/10.1016/j.xphs.2017.04.051.
[15] Fugel HJ, Nuijten M, Postma M. Stratified medicine and reimbursement issues. Front Pharmacol 2012;3:181. https://doi.org/10.3389/fphar.2012.00181.
[16] Morin K, Husereau D, Mccabe C. Personalized medicine and health care policy: from science to value. Genome Canada; 2014.
[17] Hefti E, Blanco JG. Documenting pharmacogenomic testing with CPT codes. J AHIMA 2016;87(1):56–9.
[18] Rogers L, Keeling NJ, Giri J, Gonzaludo N, Jones JS, Glogowski E, Formea CM. PARC report: a health-systems focus on reimbursement and patient access to pharmacogenomics testing. Pharmacogenomics 2020;21(11):785–96. https://doi.org/10.2217/pgs-2019-0192.
[19] Park SK, Thigpen J, Lee IJ. Coverage of pharmacogenetic tests by private health insurance companies. J Am Pharm Assoc 2020;60(2):352–356.e3. https://doi.org/10.1016/j.japh.2019.10.003.
[20] National Institute for Public Health and the Environment. Ministry of Health, Welfare and Sport., 2019, https://www.rivm.nl/en/population-screening-programmes.
[21] Australian Genomics. Funding for genetic testing to affect thousands of families., 2020, https://www.australiangenomics.org.au/funding-for-genetic-testing-to-affect-thousands-of-families/.
[22] Miller I, Ashton-Chess J, Spolders H, Fert V, Ferrara J, Kroll W, Askaa J, Larcier P, Terry PF, Bruinvels A, Huriez A. Market access challenges in the EU for high medical value diagnostic tests. Perinat Med 2011;8 (2):137–48.
[23] Vozikis A, Cooper DN, Mitropoulou C, Kambouris ME, Brand A, Dolzan V, Fortina P, Innocenti F, Lee MT, Leyens L, Macek Jr M, Al-Mulla F, Prainsack B, Squassina A, Taruscio D, van Schaik RH, Vayena E, Williams MS, Patrinos GP. Test pricing and reimbursement in genomic medicine: towards a general strategy. Public Health Genom 2016;19(6):352–63.
[24] Patrinos GP, Baker DJ, Al-Mulla F, Vasiliou V, Cooper DN. Genetic tests obtainable through pharmacies: the good, the bad and the ugly. Hum Genom 2013;7(1):17.
[25] Pavlidis C, Lanara Z, Balasopoulou A, Nebel JC, Katsila T, Patrinos GP. Meta-analysis of nutrigenomic biomarkers denotes lack of association with dietary intake and nutrient-related pathologies. OMICS 2015;19 (9):512–20.

[26] Koufaki MI, Karamperis K, Vitsa P, Vasileiou K, Patrinos GP, Mitropoulou C. Adoption of pharmacogenomic testing: a marketing perspective. Front Pharmacol 2021;12, 724311. https://doi.org/10.3389/fphar.2021.724311.

[27] Evans JP, Watson MS. Genomic testing and FDA regulation: overregulation threatens the emergence of genomic medicine. JAMA 2015;313(7):669–70.

[28] Dierking A, Schmidtke J, Matthijs G, Cassiman J-J. The EuroGentest clinical utility gene cards continued. Eur J Hum Genet 2013;21(1):1.

[29] Berwouts S, Morris M, Dequeker E. Approaches to quality management and accreditation in a genomic testing laboratory. Eur J Hum Genet 2010;18(Suppl 1):S1–19.

[30] Kampourakis K, Vayena E, Mitropoulou C, Borg J, van Schaik RH, Cooper DN, Patrinos GP. Key challenges for next generation pharmacogenomics. EMBO Rep 2014;15(5):472–6.

[31] Virelli CR, Mohiuddin AG, Kennedy JL. Barriers to clinical adoption of pharmacogenomic testing in psychiatry: a critical analysis. Transl Psychiatry 2021;11(1):509. https://doi.org/10.1038/s41398-021-01600-7.

[32] Rodríguez Vicente AE, Herrero Cervera MJ, Bernal ML, Rojas L, Peiró AM. Personalized medicine into health national services: barriers and potentialities. Drug Metab Pers Ther 2018;33(4):159–63. https://doi.org/10.1515/dmpt-2018-0017. PMID: 30391933.

[33] Rahawi S, Naik H, Blake KV, Owusu Obeng A, Wasserman RM, Seki Y, Funanage VL, Oishi K, Scott SA. Knowledge and attitudes on pharmacogenetics among pediatricians. J Hum Genet 2020;65(5):437–44. https://doi.org/10.1038/s10038-020-0723-0.

[34] Subasri M, Barrett D, Sibalija J, Bitacola L, Kim RB. Pharmacogenomic-based personalized medicine: multistakeholder perspectives on implementational drivers and barriers in the Canadian healthcare system. Clin Transl Sci 2021;14(6):2231–41. https://doi.org/10.1111/cts.13083.

[35] Rahma AT, Elbarazi I, Ali BR, Patrinos GP, Ahmed LA, Al-Maskari F. Stakeholders' interest and attitudes toward genomic medicine and pharmacogenomics implementation in the United Arab Emirates: a qualitative study. Public Health Genom 2021;24(3-4):99–109.

[36] Berwouts S, Fanning K, Morris MA, Barton DE, Dequeker E. Quality assurance practices in Europe: a survey of moleculargenomic testing laboratories. Eur J Hum Genet 2012;20(11):1118–26.

[37] Reydon TA, Kampourakis K, Patrinos GP. Genetics, genomics and society: the responsibilities of scientists for science communication and education. Perinat Med 2012;9(6):633–43.

[38] Coulter A, Jenkinson C. European patients' views on the responsiveness of health systems and healthcare providers. Eur J Pub Health 2005;15(4):355–60.

[39] Simeonidis S, Koutsilieri S, Vozikis A, Cooper DN, Mitropoulou C, Patrinos GP. Application of economic evaluation to assess feasibility for reimbursement of genomic testing as part of personalized medicine interventions. Front Pharmacol 2019;10:830.

Index

Note: Page numbers followed by *f* indicate figures, *t* indicate tables, and *b* indicate boxes.

A

Abacavir
- health economic evidence, 74–76
- *HLA-B*57:01* testing, 72–76, 74–75*t*
- major histocompatibility complex, 73
- nucleoside reverse-transcriptase inhibitor, 72–73

Acenocoumarol, 17
Adverse drug reactions (ADRs), 15–16, 23–25
- cardiovascular medications, 71–72
- classification, 71
- cost-effectiveness, 72
- economic evaluation, 72–80
- economic impacts, 72
- pharmacogenetic testing, 72
- therapeutic dose, 71

All of Us Research Program, 24, 114–115
Allopurinol, 92–93
- health economic evidence, 76, 77*t*
- immunological reaction, 76
- xanthine oxidase inhibitor, 76

Angiogenesis, 33
Antibiotics, 71–72
Anticoagulation, 91–92
Antidepressants, 92
Antiepileptic drugs (AEDs), 71–72, 90
Antiplatelet therapies, 91–92
Antipsychotic treatment, 18
Antiviral drugs, 71–72
Array genomic hybridization (AGH), 63–64
Aspirin, 17
AstraZeneca, 25
Atrial fibrillation, 102
Attributes, 57
Autism spectrum disorder (ASD), 51

B

Big data, 113
- analytics, 115
- economic evaluation of genomic tests, 114–115
- genome sequencing, 114
- sequencing tests, 115–119

Big Data to Knowledge Program (BD2K), 113–114
BigMedilytics (Big Data for Medical Analytics), 113–114
Bioaccumulation, 16–17
Bipolar disorder (BD), 18
Breast cancer, 16
Budget allocation, 140
Budget constraints, 2–4, 3*f*
Budget impact analysis, 101–102

C

Cancer
- cost-effectiveness comparator, 38
- diagnosis, 33
- methodological challenges, 35
- panel testing, 34
- pricing and reimbursement, 39–41, 40*f*, 40*b*, 42*b*
- targeted therapies, 33–34
- test and treatment inseparability, 35–36
- treatments, 33
- value of testing, 37–38

Cancer Drugs Fund (CDF), 38
Cancer therapeutics, 16–17
Candidate gene, 18
Carbamazepine (CBZ), 89–90
- cost-effectiveness, 77–78
- economic evaluations, 78, 79*t*
- first-line treatment, 76–77
- *HLA-B*15:02*, 77
- human leukocyte antigen (HLA) genotype, 79, 79–81*t*
- therapeutic recommendations, 76–77, 78*t*

Cardiovascular disease drug treatment, 17
The Center for Medical Technology Policy (CMTP), 137–138
Centers for Medicare and Medicaid Services (CMS), 137
Cervical cancer (CC) screening program, 93–94
Chemotherapeutic antitumor drugs, 33
Childhood-specific valuation methods, 59
Childhood syndromes, 141
Choice tasks, 57
Chromosomal microarrays (CMA), 47, 53–54, 60–61
Chronic myeloid leukemia (CML), 41
Clinical Laboratory Improvement Amendments (CLIA) program, 137
Clinical utility, 145
Clinical Utility Gene Cards, 142–143
Clopidogrel, 17
Colorectal cancer (CRC), 37
Commodities, 3
Comparators, 38
- economic evaluations, 52–53, 52*t*
- genomic approaches in a retrospective cohort, 54
- genomic technologies, 55

Comparators *(Continued)*
rare diseases, 55
retrospective cohort, testing sequence, 54
study designs, 55
trial cohort to compare testing methods, 53–54
Complex genomic sequencing (CGS) technology, 37
Consumption, 1–2, 4, 7
Cost-benefit analysis (CBA), 11–12
Cost components, 61
Cost-effectiveness acceptability curves (CEACs), 64
Cost-effectiveness analysis (CEA), 10, 11*f*, 47, 57, 63*f*, 77–78, 116, 139. *See also* Incremental cost-effectiveness ratio (ICER); Personalized medicine
big data, 114–115
cervical cancer screening strategies, 93
comparator, 38
cost and benefit ratio, 37, 101–102
cytology, 93–94
in genome economics model (GEM), 102–106, 104*f*
genome-guided treatments, 34
*HLA-B*15:02* genotyping, 89, 94–95, 95*f*
human papillomavirus (HPV) testing, 93–94
personalized medicine interventions in medium- and low-income countries, 94–95
pharmacogenetic dosing algorithm, 90–91
pharmacogenomic biomarkers, 90
pharmacogenomics-guided treatment, 90–91
plane, 100–101, 101*f*
targeted treatment, 35
trastuzumab, 36
treatments, 90
tumor-agnostic targeted therapies, 42
Cost-minimization analyses, 10
Costs and benefits, 2
Costs estimation, 57–58, 58*f*
Cost-utility analysis (CUA), 10, 139
Coumarinic oral anticoagulants, 17
Croatia, 102
Current Procedural Terminology (CPT), 139
Curves, indifference, 5–7, 6–7*f*
CYP2C19-guided genotyping, 91–92

D

Data analysis challenges in sequencing
counterfactual for sequencing technologies, 119
selection bias and confounding, 118
Database-driven approach, 19
Data collection challenges in sequencing
cost-effectiveness analysis, 116
N-of-1 trials in clinical medicine, 116
Data linkage, 116–117
Data management
big datasets for economic evaluation, 117–118
data linkage, 116–117
zero observations, 117
DECIPHER database, 19
Decision-making process, 139, 143
Deficient DNA mismatch repair (dMMR), 34
Demands, 2–4, 8–10, 12
Diagnostic odyssey, 48, 49*f*
cost-effectiveness, 51
next-generation sequencing (NGS), 51
in rare diseases, 67
time and healthcare resources, 47
Dihydropyrimidine dehydrogenase (DPD) enzyme, 16–17
Discrete choice experiments (DCEs), 57, 58*f*
Disease databases, 19, 20–22*t*
*DPYD*2A* biomarker, 16–17
Drug efficacy, 15, 25–26
Drug-induced toxicity, 71
Drug interactions, 71
Drug intolerance, 71
Drug rash with eosinophilia and systemic signs (DRESS), 76
Drug toxicity, 15, 18, 25–26
Dynamic efficiency, 41, 42*b*

E

Economic evaluation, 9–10
genetic and genomic tests, 88, 88*t*
in medium- and low-income countries, 88
rare diseases, 47–66
types, 10–12
Economic resources, 1–2
Economics
consumption, 1
definition, 1
distribution concept, 1
goods, services and economic resources, 1–2
health, 2–9
E-health systems, 143
Endoxifen, 16
Ethico-legal frameworks, 23
EULAC-PerMed program, 139
European Medicines Agency (EMA), 16, 74–76, 79, 81–82, 87
Euro-PGx project, 96
Exome Aggregation Consortium (ExAC), 25
Extra-welfarist approach, 10–12, 11*f*

F

Fluorescence in situ hybridization (FISH), 63–64
Fluoropyrimidine-related toxicity, 16–17
5-Fluorouracil (5-FU), 16–17
Free-market, 9–10, 144–145

G

Gastrointestinal (GI) symptoms, 72–73
General Data Protection Regulation (GDPR), 144
Generalization genome economics model (gGEM), 106–111, 107*f*
Generic model, 94–95, 95*f*
Genethics, 25–26
Genetic etiology, 48
Genetic heterogeneity, 48
Genetic mutations, 39
Genetic susceptibility, 71–72
Genetic testing
  cost of, 92–93
  countries list, 88, 88*t*
  insurance companies, 128–131
  services and genomics, 88, 127–128
Genome browsers, 19–22
Genome economics model (GEM), 102–106
  vs. classic model, 103–104, 105*f*, 106
  cost-effectiveness plane, 102–103, 104*f*
  economic depression, 102–103
  final health indicators, 105–106
  fixed-λ approach, 103–104
  generalization (gGEM), 106–109, 107*f*
  perspectives, 110–111
  quality-adjusted life-years (QALYs), 102–103
  technology assessment and evaluation, 106
Genome-guided decision-making, 4
Genome-guided treatment approaches, 34, 39, 41, 128–131
Genome sequencing, 114
  data analysis challenges
    counterfactual for sequencing technologies, 119
    selection bias and confounding, 118
  data collection challenges
    cost-effectiveness analysis, 116
    N-of-1 trials in clinical medicine, 116
  data management challenges
    big datasets for economic evaluation, 117–118
    data linkage, 116–117
    zero observations, 117
  identification, challenges, 119
Genome-wide association study (GWAS), 17–18, 77
Genomic databases
  categories, 19
  database-driven approach, 19
Genomic diagnosis, 64*b*
Genomic medicine, 128, 129*f*
  contribution, 132
  cost-effectiveness, 101–102
  health economics, 100
  media and press, 131
  medical conditions, 99
  personalized medicine, 102
  pharmacogenomic (PGx)-guided warfarin treatment, 102
  strategic planning, 125
Genomics education, 143–144
Genomic testing, 12, 34, 59, 65, 140–141
  clinical and healthcare system implications, 50–51
  coverage, pricing, and reimbursement strategies, 138–139
  diagnostic testing process, 48
  diseases and phenotypes, 144
Genomic variants, 15, 18, 20–22*t*, 118
Global population, 48
Goods, 1–2
Greece
  financial crisis, 128–131
  medium opposition, 128–131
  opportunities and threats in genomics, 131–132
  preliminary assessment, 127–131, 129*f*, 130*t*
Greek Ministry of Health and the public health insurance funds, 128–132
Gross domestic product (GDP), 92

H

Han Chinese ancestry, 18
Healthcare budget, 7
Healthcare providers, 8–10
Healthcare Resource Group (HRG), 117–118
Healthcare system, 2–3, 9, 87
Health economics, 2–9
  budget constraints and demand, 2–4
  indifference curves, 5–7, 6–7*f*
  social welfare, 7–9
  utility, 4–5, 5*f*
Health professionals, 23–24
Health-related quality of life (HRQoL), 37
Health services, 3
Health technology agencies (HTAs), 136–139, 145
Health technology assessment (HTA), 35–36, 38, 87, 89
Hemoglobin variants (HbVar), 19
Hereditary cancers, 135
High-income economies challenges
  low-income and lower middle-income countries outside Europe, 89
  upper middle-income countries in Europe, 89–91
Highly active antiretroviral therapy (HAART), 72–73
Highly specialized technologies program (HSTP), 36
HIV-positive patients, 74–76, 74–75*t*
*HLA-B*15:02* genotyping, 94–95, 95*f*
*HLA-B*57:01* screening, 72–76, 74–75*t*
*HLA-B*58:01* screening, 76, 77*t*, 92–93
Hospital Episode Statistics (HES), 114–115
Human Genome Organization-Mutation Database Initiative (HUGO-MDI), 19

Human Genome Variation Society (HGVS), 19
Human Heredity and Health in Africa (H3Africa) study, 25, 96
Human leukocyte antigen (HLA)
abacavir-induced hypersensitivity reaction, 72–76
allopurinol-induced hypersensitivity reaction, 76
antigens, 71–72
carbamazepine-induced hypersensitivity reaction, 76–80, 78–81*t*
Human papillomavirus (HPV) testing, 93–94
Hypersensitivity drug reactions (HDRs), 71
Hypersensitivity reaction (HSR), 71–72
Hypersensitivity syndrome, 76–77
Hypothetical goods, 3

I

Idiopathic developmental disability (IDD), 57, 60–61
Idiosyncratic reactions, 71
Incremental cost-effectiveness ratio (ICER), 10–11, 39, 62, 64–65, 92, 100
pharmacogenomic (PGx)-guided warfarin treatment, 102
technical matter, 101
willingness to pay (wtp/λ), 100–101
Indication-based pricing (IBP), 39
Indifference curves, 5–7, 6–7*f*
Innovation, 142, 145, 146*f*
defined, 102
healthcare system, 101
reimbursement, 105
willingness to pay (wtp/λ), 109
Insurance funds, 4
International Normalized Ratio (INR), 90–91
The International Society for Pharmacoeconomics and Outcomes Research (ISPOR), 36
Invasive cervical cancer (ICC), 94
IPREDICT Melbourne Genomics Health Alliance study, 37
Irinotecan, 16
Iron triangle, 140

K

Kaldor-Hicks criterion, 8

L

Launch sequence, 39, 40*b*
Literature mining, 19
Lithium treatment, 18
Locus/disease/ethnic/other-specific databases, 19, 20–22*t*
Long-QT syndrome (LQTS), 65

M

Maculopapular exanthema (MPE), 76–77
Marginal utility, 4, 5*f*
Market power, 9–10
Massively parallel sequencing (MPS), 37
Master protocols, 38
Medium- and low-income economies
allopurinol, 92–93
anticoagulation and antiplatelet therapies, 91–92
antidepressants, 92
cost-effectiveness of personalized medicine, 94–95, 95*f*
human papillomavirus (HPV) testing, 93–94
Mexico National Institute of Genomic Medicine (INMEGEN), 96
Microsatellite instability-high (MSI-H), 34
Model-based cost-effectiveness analysis, 146
Monetary cost, 99
Monetary units, 2–3
Monoclonal antibodies (MABs), 33

N

National Health Service (NHS), 10, 60–61, 137
National Human Genome Research Institute (NHGRI), 137–138
The National Institute for Health and Care Excellence (NICE), 36
Natural language processing (NLP), 19
Negative predictive value (NPV), 77
Next-generation sequencing (NGS), 22, 25–26, 47, 131, 144, 146
cost-effectiveness of, 51
diagnostic odyssey, 47, 50*f*
in rare diseases, 52, 52*t*
Next-generation sequencing (NGS)-based gene panel tests, 34, 41–42
Non-small cell lung cancers (NSCLCs), 34
Nonsteroidal anti-inflammatory drugs (NSAIDs), 71–72

O

Oncology, 33–35, 42
Oncotype Dx, 41
One-size-fits-all approaches, 145–146
Online Mendelian Inheritance in Man (OMIM) database, 19
Opportunity cost, 3

P

Pan American Health Organization/World Health Organization (PAHO/WHO), 138
Pareto improvement concept, 8
Patient stigmatization, 132
Pembrolizumab, 34
Performance ratio (PR), 109
Personalized healthcare, 2
Personalized medicine, 102
cost-effectiveness, 94–95, 95*f*

coverage, pricing, and reimbursement strategies, 138–139
drug treatment, 135
economic and policymaking issues, 135–136
economic evaluation, 24
ethical and legal issues, 23
ethical, legal, and societal (ELSI) issues, 123
genomics knowledge, 23–24
high-income countries
low-income and lower middle-income countries outside Europe, 89
upper middle-income countries in Europe, 89–91
institutions involved in pricing and reimbursement, 137–138
interdependencies, 136, 136*f*
interventions, 24
large-scale nationwide efforts, 24–25
in medium- and low-income countries, 91–94
monogenic disorders, 135
patient-specific treatment decisions, 123
policymaking environment in Greece, 131–132
pre-emptive genotyping, 24
proposed strategy for pricing and reimbursement, 140–144
public healthcare system, 136
stakeholders
analysis, 127–131
views and opinions, 124–126
Personal utility, 57–58, 58*b*
Pharmacodynamic-associated genes, 18
Pharmacogenes, 22
Pharmacogenomic biomarkers
antipsychotic treatment, 19
cancer therapeutics, 16–17
cardiovascular diseases drug treatment, 17
pharmacogenes, 22
psychiatric diseases, 18–19
US Food and Drug Administration (FDA)-approved, 18
Pharmacogenomics (PGx), 140
clinical care, 16–17
cost-effectiveness, 24, 128–131
databases, 19, 20–22*t*
disease heterogeneity and genetic complexity, 15
electronic tools, 19–22, 20–22*t*
genomic information, 87
genomic medicine, 129*f*
medications, 15
public health genomics-related issues, 15
therapy, 102
Pharmacotherapy, 24–25
Phenprocoumon, 17
Phenytoin (PHT), 90
PolicyMaker tool
design features, 125
genomic and personalized medicine, 127–128
policy information, 124–125
political mapping tool, 125
schematic drawing, 125, 126*f*
stakeholder views and opinions, 124–125
strategic planning for policy formulation, 125
Policymaking, 124–125, 128, 135–136
Politics, 1
Population databases, 19, 20–22*t*
Portable legal consent (PLC), 23
Positive predictive value (PPV), 77
Pricing, 39–41, 40*f*, 40*b*, 42*b*
acceptable and affordable prices for healthcare system, 140–142
clinical utility/actionability of testing outcomes, 142–143
coverage, 138–139
genomic testing services, 138–139
genomic tests and information by physicians, 142–143
innovation, 142
institutions, 137–138
interdependencies, 136, 136*f*
novel and existing diagnostic procedures, and monitor patient safety, 143–144
patient needs and clinical utility/actionability of testing outcomes, 142–143
proposed strategy, 140–144
and reimbursement, 137–144
research of personalized medicine investment, 143–144
Progression-free survival (PFS), 99, 108
Prospective, block-randomized, controlled clinical study (PREPARE), 24–25
Psychiatric diseases, 18–19
Psychiatric Genetics Consortium (PGC), 25
Public healthcare system, 23–24, 136

Q

Qatar Genome Project, 96
Quality-adjusted life-week (QALW), 92
Quality-adjusted life-years (QALYs), 10, 12, 34, 37, 41, 51, 55–56, 90–91, 93–94, 99–101, 108, 110

R

Randomized controlled trials (RCTs), 52–53, 113–114, 119
Rare diseases, 116
comparators and study design, 52–55, 52*t*, 53*f*
recommendations, 55
selected literature examples, 53–54
diagnostic odyssey, 48, 49*f*
estimate costs
economic evaluations, 59, 59*t*
recommendations, 61–62

Rare diseases *(Continued)*
estimate relationship between costs and outcomes, 62–66, 62*f*, 64*b*
cost-effectiveness, net benefit, and personal utility, 63–65
cost per quality adjusted life years (QALYs) among relatives of affected individuals, 65
costs and consequences of technology adoption, 65–66
recommendations, 66
genetic origin, 47
genomics opportunity
clinical and healthcare system implications, 50–51
genomic testing, 48
health and non-health outcomes, 55–59, 55*t*, 56*f*
cases avoided through testing, 57, 58*b*
cost per additional diagnosis and changes in management, 56–57
personal utility, 57–58, 58*f*
recommendations, 59
RedETSA, 138
Reimbursement, 39–41, 40*f*, 40*b*, 42*b*, 105, 108–109
clinical utility/actionability of testing outcomes, 142–143
coverage, pricing, 138–139
efficacy, quality, and fairness, 142
genomic testing services, essential, 136, 140–142
genomic tests and information by physicians, 142–143
institutions, 137–138
investment, 143–144
pricing and, 136, 136*f*
Resource allocation, 107*f*, 111
Resource-rich healthcare system, 2–3
Risperidone, 90
Rituximab, 33

S

Safety assurance, 142
Safety net, 9
Secondary findings (SFs), 51
Secretary's Advisory Committee on Genetics, Health, and Society (SACGHS), 137
Sequence databases, 19, 20–22*t*
Sequencing tests challenges
data analysis
counterfactual for sequencing technologies, 119
selection bias and confounding, 118
data collection
cost-effectiveness analysis, 116
N-of-1 trials in clinical medicine, 116
data management
big datasets for economic evaluation, 117–118
data linkage, 116–117
zero observations, 117
identification, challenges, 119
Severe cutaneous adverse drug reactions (SCARs), 71–72
Side effects, 71
Single-gene tests, 48, 51
SN-38 glucuronidation, 16
Social activities, 1, 4, 7–8
Social fairness, 101
Social welfare, 7–9
Societal aspects, 22–24
Societal choices, 101
Societal preference, 101
Stakeholders, 116–117, 135–136, 141, 145
eliciting and analyzing views and opinions of, 124–126, 126–127*f*
genomic and personalized medicine in Greece, 127–131, 129*f*, 130*t*
identification, personalized medicine, 124
opportunities and threats, 131–132
position and feasibility graphs, 125, 127*f*
Statins, 17
Stevens-Johnson syndrome/toxic epidermal necrolysis (SJS/TEN) in patients, 71–72, 76–77, 79, 89–90

T

Tamoxifen, 16
Targeted therapy
challenges, 35, 38
cost-effectiveness, 35
genetic mutations, 39
genomic alterations, 38
genomic tests, 34
oncology drug, 33
target genetic mutations, 39
Technology appraisal program (TAP), 36
Testing values, 37–38
Time to progression (TTP), 99
Toxic Epidermal Necrolysis (TEN), 89–90
Trastuzumab, 36, 39
Trial designs, 38

U

Ubiquitous Pharmacogenomics Consortium (U-PGx), 24–25
*UGT1A1*28* allele, 16
US Department of Health and Human Services (HHS), 74–76
United States Food and Drug Administration (FDA), 16, 34, 74–76, 79, 81–82, 87, 139
Utility, 4–5, 5*f*, 92

V

Value-based pricing (VBP), 39, 145
Value of testing, 37–38
Variants of unknown significance (VUS), 51

W

Warfarin, 17, 91
Welfarism
  economic evaluation, 10–12
  extra-welfarist approach, 10–12, 11*f*
Welfarist approach, 11–12
Whole exome studies (WES), 19, 48, 51, 54, 65–66
Whole-genome sequencing (WGS) data, 114–115
Whole genome studies (WGS), 19, 48, 61
Willingness to pay (wtp/λ), 57–58, 66, 100–101, 101*f*, 103–104, 110–111

X

Xenobiotics, 15

Z

Zero observations, 117

Printed in the United States
by Baker & Taylor Publisher Services